A Text Book for F.Y.B.Sc. (Computer Science)
Semester - II
MATHEMATICS PAPER - II : MTC-122 (Credit 2)
Choice Based Credit System (CBCS) (2019 Pattern)

GRAPH THEORY

Dr. Kalyanrao Takale
M.Sc., B.Ed., Ph.D.
RNC Arts, JDB Commerce and NSC Science College
Nashik Road, Nashik

Dr. Shrikisan Gaikwad
M.Sc., B.Ed., M.Phil., Ph.D.
New Arts, Commerce and Science College
Ahmednagar

Dr. Mrs. Nivedita Mahajan
M.Sc., B.Ed., Ph.D.
Modern College of Arts, Science and Commerce
Shivajinagar, Pune

Dr. Amjad Shaikh
M.Sc., Ph.D.
AKI's, Poona College of Arts, Science and Commerce
Pune

Prof. Mrs. Shamal Deshmukh
M.Sc., M.Phil.
Modern College of Arts, Science and Commerce
Ganeshkhind, Pune

Prof. S. R. Patil
M.Sc.
Ex. HOD., S.M. Joshi College
Hadapsar, Pune

N5065

GRAPH THEORY ISBN 978-93-89686-60-9

First Edition	:	**December 2019**
©	:	**Authors**

Published By :
NIRALI PRAKASHAN
Abhyudaya Pragati, 1312, Shivaji Nagar
Off J.M. Road, PUNE – 411005
Tel - (020) 25512336/37/39, Fax - (020) 25511379
Email : niralipune@pragationline.com

➢ **DISTRIBUTION CENTRES**

PUNE

Nirali Prakashan : 119, Budhwar Peth, Jogeshwari Mandir Lane, Pune 411002, Maharashtra
(For orders within Pune) Tel : (020) 2445 2044, Mobile : 9657703145
Email : niralilocal@pragationline.com

Nirali Prakashan : S. No. 28/27, Dhayari, Near Asian College Pune 411041
(For orders outside Pune) Tel : (020) 24690204, Mobile : 9657703143
Email : bookorder@pragationline.com

MUMBAI

Nirali Prakashan : 385, S.V.P. Road, Rasdhara Co-op. Hsg. Society Ltd.,
Girgaum, Mumbai 400004, Maharashtra; Mobile : 9320129587
Tel : (022) 2385 6339 / 2386 9976, Fax : (022) 2386 9976
Email : niralimumbai@pragationline.com

➢ **DISTRIBUTION BRANCHES**

JALGAON

Nirali Prakashan : 34, V. V. Golani Market, Navi Peth, Jalgaon 425001, Maharashtra,
Tel : (0257) 222 0395, Mob : 94234 91860; Email : niralijalgaon@pragationline.com

KOLHAPUR

Nirali Prakashan : New Mahadvar Road, Kedar Plaza, 1st Floor Opp. IDBI Bank, Kolhapur 416 012
Maharashtra. Mob : 9850046155; Email : niralikolhapur@pragationline.com

NAGPUR

Nirali Prakashan : Above Maratha Mandir, Shop No. 3, First Floor,
Rani Jhanshi Square, Sitabuldi, Nagpur 440012, Maharashtra
Tel : (0712) 254 7129; Email : niralinagpur@pragationline.com

DELHI

Nirali Prakashan : 4593/15, Basement, Agarwal Lane, Ansari Road, Daryaganj
Near Times of India Building, New Delhi 110002 Mob : 08505972553
Email : niralidelhi@pragationline.com

BENGALURU

Nirali Prakashan : Maitri Ground Floor, Jaya Apartments, No. 99, 6th Cross, 6th Main,
Malleswaram, Bengaluru 560003, Karnataka; Mob : 9449043034
Email: niralibangalore@pragationline.com

Other Branches : Gujarat, Hyderabad, Chennai, Kolkata

niralipune@pragationline.com | www.pragationline.com

Also find us on f www.facebook.com/niralibooks

Preface

This book is based on a course Graph theory. We write this book as per the revised syllabus of F.Y. B.Sc.(Computer Science) Mathematics, revised by Savitribai Phule Pune University, Pune, implemented from June 2019. Graph theory is the most useful subject in all branches of mathematics and it is used extensively in applied mathematics and engineering.

Graphs theory is the study of graphs, which are mathematical structures used to model pairwise relations between objects. It is a bridge connecting mathematics with various branches of computer science. We study how problems in almost every conceivable discipline can be solved using graph models.

The aim of this textbook on Graph theory is to introduce basic concept in graph theory and some shortest path algorithm to model computation-related problems..

In chapter 1, basic concepts and terms of graph theory have been introduced. We will also introduce several important families of graphs often used as examples and in models. We will show how to represent graphs in several different ways and a one-to-one correspondence between their vertex sets that preserves edges.

In the second chapter, we discuss concepts like walk, path, cycle etc. which would help us to understand the definition of connected graph. We also study isthmus and cut vertex and algorithm to find shortest path between two vertices.

The aim of third chapter is to equip students with the Euler and Hamilton circuits. The chapter concludes with their application to Travelling salesman and Chinese Postman problem.

In fourth chapter, we will focus on a particular type of graph called a tree, so named because such graphs resemble trees. Trees are used as models in such diverse areas as computer science, chemistry, geology, botany, and psychology. We will describe a variety of such models based on trees.

We are very much thankful to **Mr. Dinesh Furia** and **Mr. Jignesh Furia**, Nirali Prakashan, Pune, for valuable cooperation and guidance. Also, we are specially thankful to Mrs. Nanda Takale for her valuable cooperation. All authors devote this book to their parents.

We welcome any opinions and suggestions which will improve the future editions and help readers in future.

In case of queries/suggestions, send an email to: **kalyanraotakale@rediffmail.com**

Syllabus PAPER-II: MTC-122: Graph Theory

1. An Introduction to graph [10]

1.1 Definitions, Basic terminologies and properties of graph, Graph models.

1.2 Special types of graphs, basic terminologies, properties and examples of directed graphs .Types of diagraphs.

1.3 Some applications of special types of graph.

1.4 Matrix representation and elementary results, Isomorphism of graphs.

2. Connected graph [08]

2.1 Walk, trail, path, cycle, elementary properties of connectedness. Counting paths between vertices (by Warshalls algorithm).

2.2 Cut edge (Bridge), Cut vertex, cut set, vertex connectivity, edge connectivity, and Properties.

2.3 Shortest path problem, Dijkstras algorithm.

3. Euler and Hamilton path [07]

3.1 The Konigsberg bridge problem, Euler trail, path, circuit and tour, elementary properties and Fleurys algorithm.

3.2 Hamilton path, circuit, elementary properties and examples.

3.3 Introduction of Travelling salesman problem, Chinese postman problem.

4. Trees [10]

4.1 Definitions, basic terminologies, properties and applications of trees.

4.2 Weighted graph, definition and properties of spanning tree, shortest spanning tree, Kruskals algorithm, Prims algorithm.

4.3 M-ary tree, binary tree, definitions and properties, tree traversal: preorder, inorder, postorder, infix, prefix, postfix notations and examples.

Contents

1 An Introduction to Graph   **1**

   1.1 Graph Terminology 1

      1.1.1 Basic Definitions 1

      1.1.2 Degree of a vertex 3

   1.2 Graph Models 5

   1.3 Types of a Graph 7

   1.4 Types of Digraphs 10

   1.5   Representing Graphs and Graph Isomorphism 19

      1.5.1   Representation of Graphs 19

      1.5.2   Isomorphism of Graphs 22

   1.6 Operations on Graphs 32

      1.6.1   Subgraphs 32

      1.6.2   Deletion of vertices and edges from a Graph 34

      1.6.3   Complement of a Graph 36

      1.6.4   Union, Intersection, Ring sum and Product of Graphs 37

2 Connected graphs   **49**

   2.1 Walk, Path and Circuits 49

   2.2 Connected Graph 51

   2.3 Counting paths between vertices 53

   2.4 Isthmus and Cut vertex 59

      2.4.1   Connectivity 60

   2.5 A Shortest-Path algorithm 66

3 Euler and Hamilton Path   **75**

   3.1 Euler Path and Euler circuit 75

      3.1.1   Fleury's algorithm 78

   3.2 Hamilton Path and Circuit 85

4 Trees   94

4.1   Definition and Properties of a tree   94

4.2   Centre of a tree   96

4.3   Spanning trees: 98

    4.3.1   Shortest spanning tree, Kruskal's and Prims algorithm 102

4.4   Binary tree 111

4.5   m-ary trees: 113

4.6   Tree Traversal 117

    4.6.1   Infix, Prefix and Postfix notation 123

Chapter 1

An Introduction to Graph

Introduction

Graph theory is intimately related to many branches of mathematics. It is widely applied in subjects like, Computer Technology, Communication Science, Electrical Engineering, Physics, Architecture, Operations Research, Economics, Sociology, Genetics, etc. In the earlier stages it was called slum Topology. Euler, Cayley, Sir William Hamilton, Lewin, and Kirchoff laid foundations to the graph theory. Graph theory was born in 1736 with Euler's paper on Konigsberg bridge problem. The Konigsberg bridge problem is the best known example in graph theory. It was a long pending problem. Euler solved this problem by means of a graph. Euler became father of graph theory. Kirchoff, Cayley, Mobius, Hamilton and De Morgan have laid strong foundations and contributed much to the development of the subject. In this chapter, basic concepts and terms of graph theory have been introduced.

1.1 Graph Terminology

1.1.1 Basic Definitions

Graphs are discrete structures consisting of vertices and edges that connect these vertices. There are different kinds of graphs, depending on whether edges have directions, whether multiple edges can connect the same pair of vertices, and whether loops are allowed. Problems in almost every conceivable discipline can be solved using graph models.

Definition 1.1. *A graph $G = (V, E)$ consists of two sets V and E where, V is a nonempty set of vertices (or nodes) and E is a set of edges. Each edge $e \in E$ is associated with an unordered pair (v_1, v_2) of vertices $v_1, v_2 \in V$.*

For example in the following graph $V = \{v_1, v_2, v_3, v_4, v_5, v_6\}$, $E = \{e_1, e_2, \cdots, e_{10}\}$

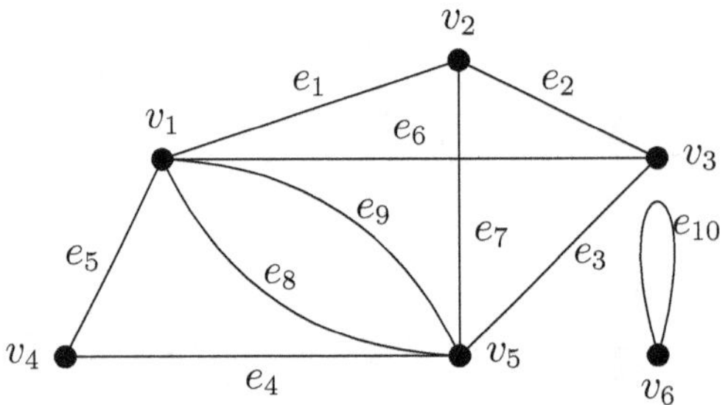

Figure 1.1

Note that edge e_4 corresponds to unordered pair (v_4, v_5) Also edge e_{10} corresponds to (v_6, v_6). Such that edge whose both endvertices are same is called as **self loop.**

Definition 1.2. *A **directed graph (or digraph)** G(V ,E) consists of a nonempty set of vertices V and a set of directed edges (or arcs) E. Each directed edge is associated with an ordered pair of vertices. The directed edge associated with the ordered pair (u, v) is said to start at u and end at v.*

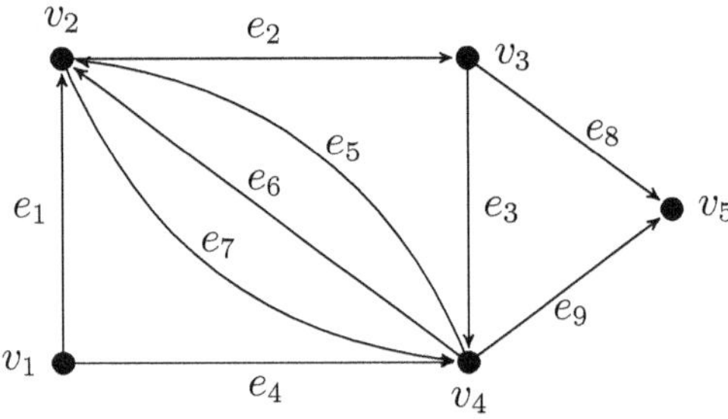

Figure 1.2

When we depict a directed graph with a line drawing, we use an arrow pointing from u to v to indicate the direction of an edge that starts at u and ends at v. A directed graph may contain loops and it may contain multiple directed edges that start and end at the same vertices. A directed graph may also contain directed edges that connect vertices u and v in both directions;

We introduce some of the basic vocabulary of graph theory in this section. We will use this vocabulary later in this chapter when we solve many different types of problems.

Definition 1.3. *Two vertices u and v in an undirected graph G are called **adjacent** (or neighbors) in G if u and v are endpoints of an edge e of G. Such an edge e is called **incident** with the vertices u and v and e is said to connect u and v.*

1.1.2   Degree of a vertex

Definition 1.4. *The degree of a vertex v in an undirected graph G(V, E) is the number of edges incident with it and is denoted by d(v).*

Remark 1.1. *Every edge of the graph G contributes two degrees, one at each end vertex. Moreover, the loop at the vertex v contributes two degrees at v.*

Remark 1.2. *A vertex of degree zero is called **isolated**.*

Remark 1.3. *A vertex is **pendant** if and only if it has degree one. Consequently, a pendant vertex is adjacent to exactly one other vertex.*

Example 1.1. *Find the number of vertices, the number of edges, and the degree of each vertex in the given undirected graph. Identify all isolated and pendant vertices.*

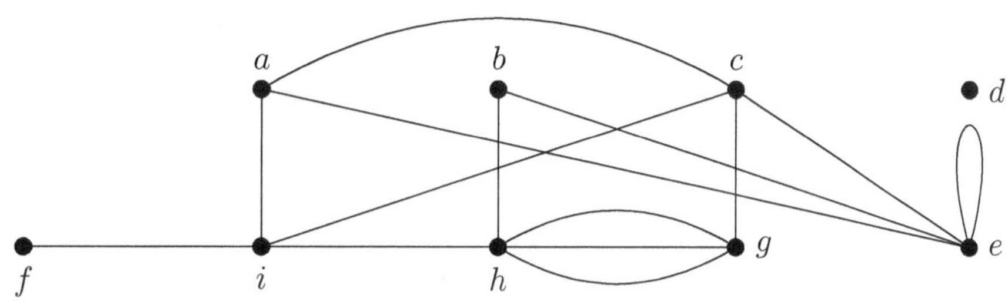

Figure 1.3

Solution: There are 9 vertices here, and 13 edges. The degree of each vertex is the number of edges incident to it. Thus deg(a) = 3, deg(b) = 2, deg(c) = 4, deg(d) = 0 (hence d is isolated), deg(e) = 5, deg(f) = 1(hence d is pendant), deg(g) = 4, deg(h) = 5, and deg(i) = 4.

Theorem 1.1 (The Handshaking Lemma).

Let G = (V ,E) be an undirected graph with m edges. Then

$$\sum_{v \in V} deg(v) = 2m.$$

That is, total degree of the graph is two times number of edges in the graph.

Remark 1.4. *This result is applicable even if multiple edges and loops are present in a graph $G(V, E)$.*

Theorem 1.2. *An undirected graph $G(V, E)$ has an even number of vertices of odd degree.*

Proof: Let V_1 and V_2 be the set of vertices of even degree and the set of vertices of odd degree, respectively, in an undirected graph $G = (V, E)$ with m edges. Then by Handshaking lemma, total degree of G is $2m$. That is

$$\sum deg(v) = 2m.$$

We separate the sum on left hand side into two parts. In the first part we take vertices of even degree and in the second part the vertices of odd degree.

$$\therefore \quad \sum_{v \in V_1} deg(v) + \sum_{v \in V_2} deg(v) = 2m.$$

The first term in the left-hand side of the above equation is even. Furthermore, the sum of the two terms on the left-hand side is even, as this sum is 2m. Hence, the second term in the sum is also even. Because all the terms in this sum are odd, there must be an even number of such terms. Thus, there are an even number of vertices of odd degree.

- **Degree of a vertex in a directed graph**

Definition 1.5. *When (u, v) is an edge of the graph G with directed edges, u is said to be **adjacent** to v and v is said to be adjacent from u. The vertex u is called the **initial vertex** of (u, v), and v is called the **terminal or end vertex** of (u, v). The initial vertex and terminal vertex of a loop are the same.*

Definition 1.6. *let v be a vertex in a digraph G. The **in-degree** of a vertex v, denoted by $deg^-(v)$, is the number of edges with v as their terminal vertex. The **out-degree** of v, denoted by $deg^+(v)$, is the number of edges with v as their initial vertex.*

Remark 1.5. *A loop at a vertex contributes 1 to both the in-degree and the out-degree of this vertex.*

Example 1.2. *Find the in-degree and out-degree of each vertex in the digraph G with directed edges shown in following figure.*

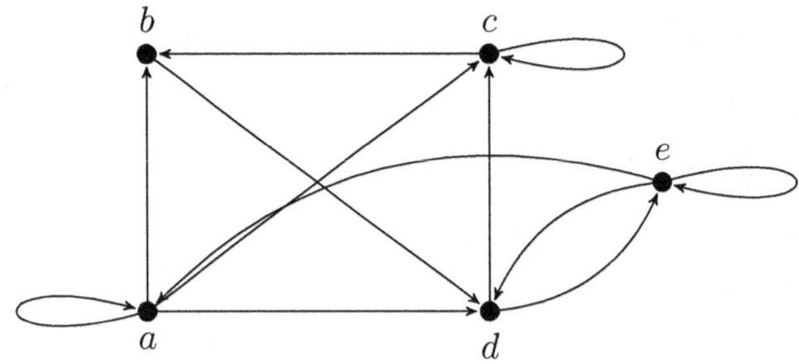

Figure 1.4: **The Directed Graph G.**

Solution:

$$deg^-(a) = 2 \qquad deg^+(a) = 4$$
$$deg^-(b) = 2 \qquad deg^+(b) = 1$$
$$deg^-(c) = 3 \qquad deg^+(c) = 2$$
$$deg^-(d) = 3 \qquad deg^+(d) = 2$$

Theorem 1.3. *Let $D = (V, E)$ be a directed graph. Then*

$$\sum_{v \in V} d^+(v) = \sum_{v \in V} d^-(v) = |E|.$$

Remark 1.6. *If we ignore the direction of an edge, we get edges of a usual graph. The resulting graph is called the **underlying graph** of the digraph D.*

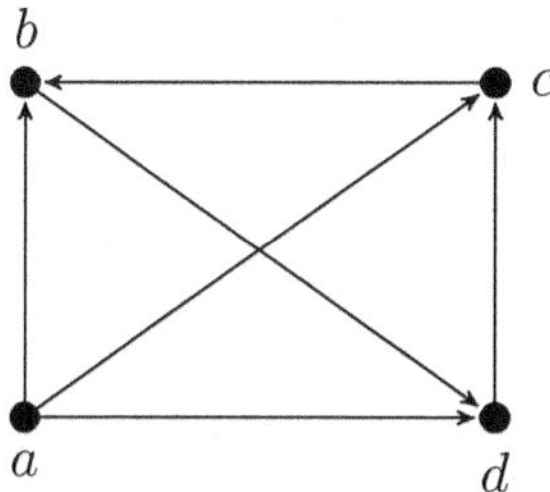

Figure 1.5: **The Directed Graph G.**

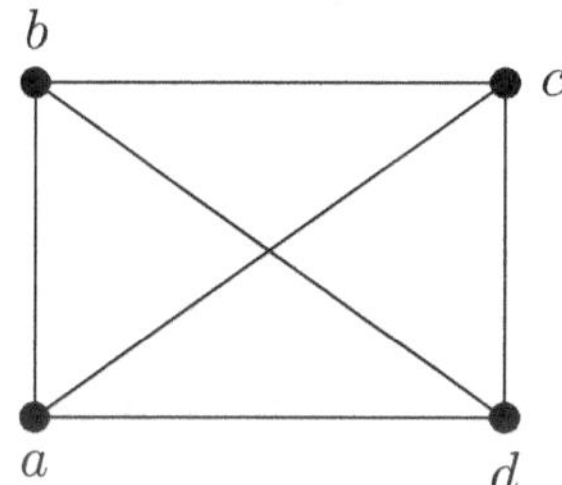

Figure 1.6: **The Underlying Graph G.**

1.2   Graph Models

Graphs are used in a wide variety of models. We present here a few graph models from diverse fields

1. **Acquaintanceship Graphs**: We can use a simple graph to represent whether two people know each other, that is, whether they are acquainted. Each person in a particular group of people is represented by a vertex. An undirected edge is used to connect two people when these people know each other, when we are concerned only with acquaintanceship. The acquaintanceship graph of all people in the world has more than six billion vertices and probably more than one trillion edges.

   - **Illustrative example:**

   Draw the acquaintanceship graph that represents the Tom and Patricia, Tom and Hope, Tom and Sandy, Tom and Amy, Tom and Marika, Jeff and Patricia, Jeff and Marika and Amy and

Marika know each other but none of the other pairs of people listed know each other.

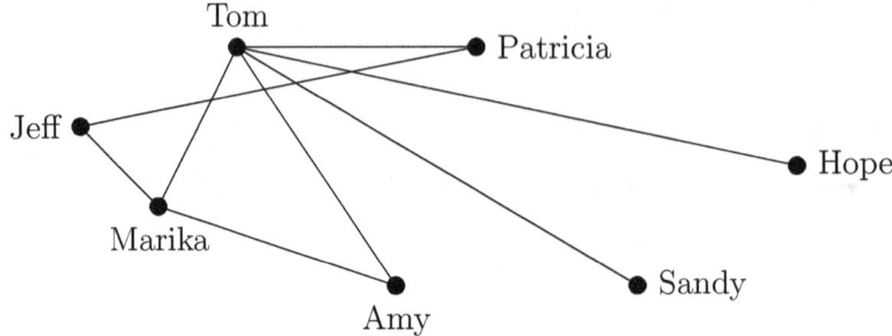

2. **Road Maps**: Graphs can be used to model road networks. In such models, vertices represent intersections and edges represent roads. To build such models, we use undirected edges to represent two-way roads and we use directed edges to represent one-way roads. Multiple undirected edges represent multiple two-way roads connecting the same two intersections. Multiple directed edges represent multiple one-way roads that start at one intersection and end at a second intersection. Loops represent loop roads.

3. **The Web Graph**: The World Wide Web can be modeled as a directed graph where each Web page is represented by a vertex and where an edge starts at the Web page a and ends at the Web page b if there is a link on a pointing to b. Because new Web pages are created and others removed somewhere on the Web almost every second, the Web graph changes on an almost continual basis. Many people are studying the properties of the Web graph to better understand the nature of the Web. The Web graph is used by the Web crawlers that search engines use to create indexes of Web pages.

4. **Niche Overlap Graphs in Ecology**: The competition between species in an ecosystem can be modeled using a niche overlap graph. Each species is represented by a vertex. An undirected edge connects two vertices if the two species represented by these vertices compete (that is, some of the food resources they use are the same). A niche overlap graph is a simple graph because no loops or multiple edges are needed in this model.

   • **Illustrative example:**

Construct a Niche Overlap Graph for the six species of birds where the hermit thrush competes with the robin and with the blue jay, the robin also competes with the mocking bird, the mocking bird also competes with the blue jay and the nuthatch competes with the hairy woodpecker.

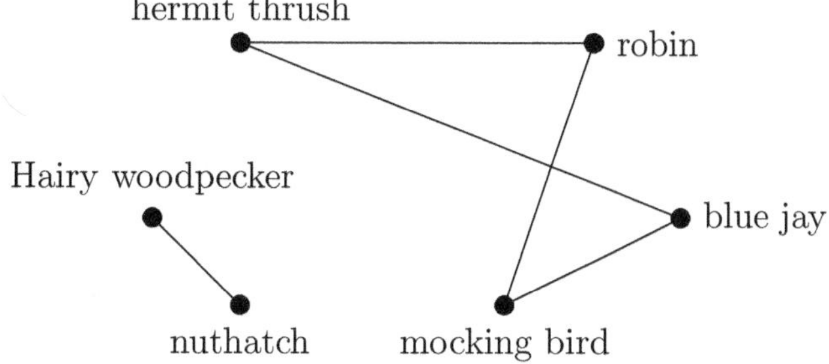

1.3 Types of a Graph

We will nowintroduce several classes of simple graphs. These graphs are often used as examples and arise in many applications.

1. Simple Graph

A graph G with no self-loops and multiple edges is called a simple graph.

The graphs in following figures are simple graphs.

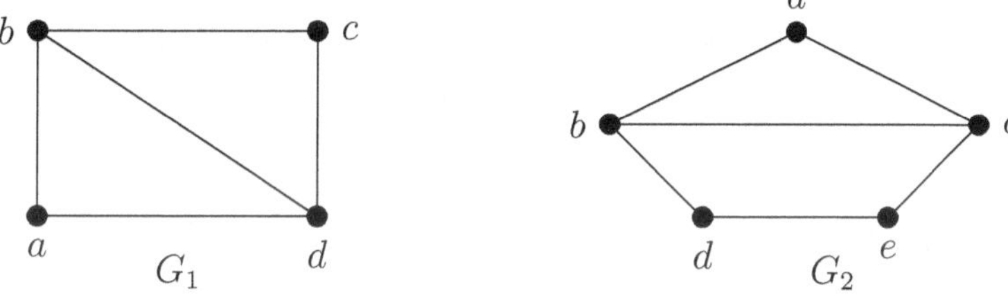

Figure 1.7

2. Null Graph

A graph G = (V, E) in which the set of edges E is empty is called a Null graph. (see Fig. 1.8)

Figure 1.8

3. Weighted Graph

A graph G is in which weight is assigned to every edge is called a weighted graph.

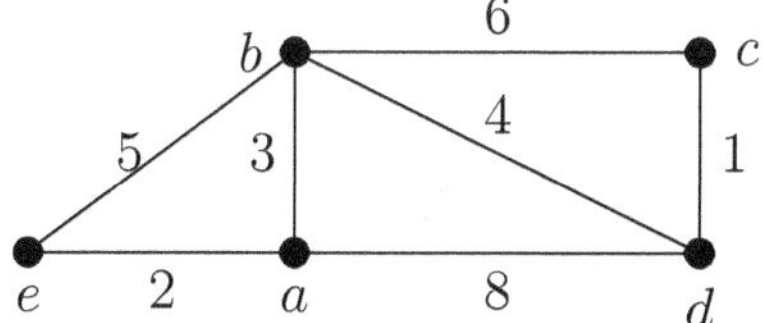

4. Complete Graph

A complete graph on n vertices, denoted by K_n, is a simple graph that contains exactly one edge between each pair of distinct vertices.

The graphs K_n, for $n = 1, 2, 3, 4, 5$ are displayed in Fig. 1.10

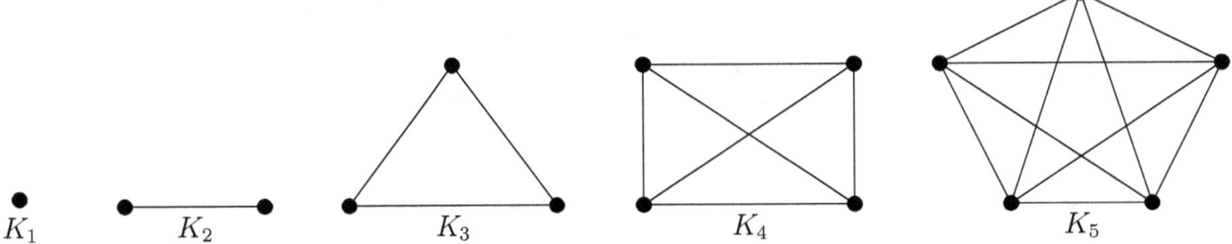

Figure 1.10

Remark 1.7.   *i. In a complete graph there is an edge between every pair of vertices.*

ii. A complete graph K_n has exactly $\dfrac{n(n-1)}{2}$ edges.

5. Cycles

A cycle C_n, $n \geq 3$, consists of n vertices $v_1, v_2, \cdots, v_n$ and edges $\{v_1, v_2\}, \{v_2, v_3\}, ..., \{v_{n-1}, v_n\}$, and $\{v_n, v_1\}$. The cycles C_3, $C_4, C_5,$ and C_6 are displayed in Figure 1.11.

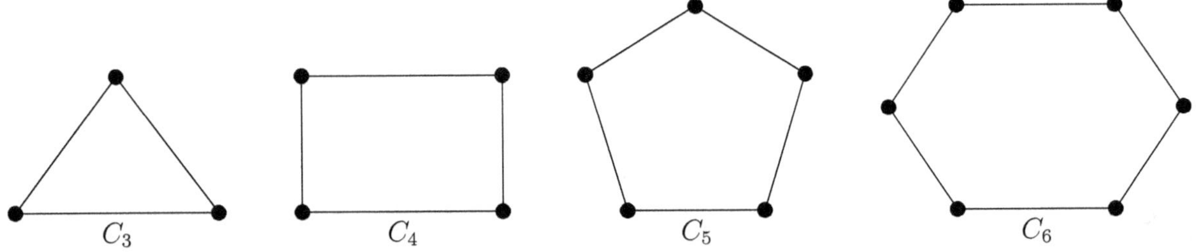

Figure 1.11

6. Wheels

We obtain a wheel W_n when we add an additional vertex to a cycle C_n, for $n \geq 3$, and connect this new vertex to each of the n vertices in C_n, by new edges. The wheels $W_3, W_4, W_5,$ and W_6 are displayed in Following Figure.

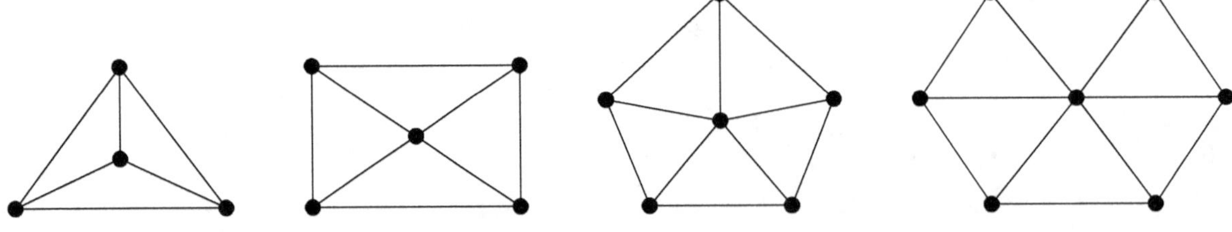

7. n-Cubes

The $n-$cube denoted by Q_n, is the graph that has vertices representing the 2^n bit strings of length n. Two vertices are adjacent if and only if the bit strings that they represent differ in exactly one bit position. The graphs Q_1, Q_2, Q_3 are displayed in Figure 1.12. Thus Q_n has 2^n vertices and $n \cdot 2^{n-1}$ edges also it is regular of degree n.

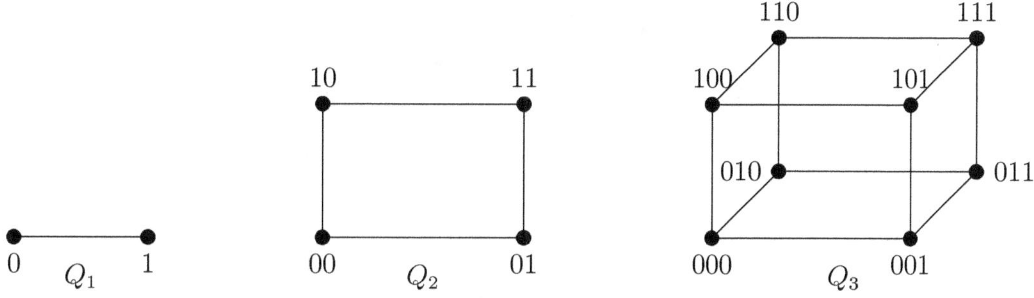

Figure 1.12

8. K-regular Graph

A Graph G is said to be k-regular, if every vertex of G has degree k.
A graph G given below is 2-regular graph.

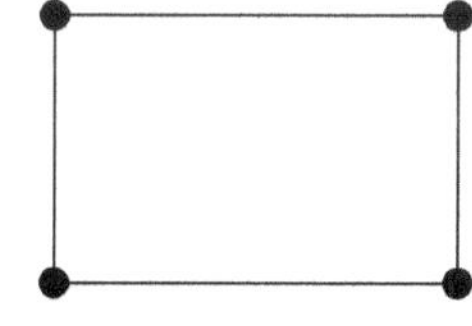

Figure 1.13

Remark 1.8. *If G is a regular graph of degree 3, it is called a* **cubic graph.**

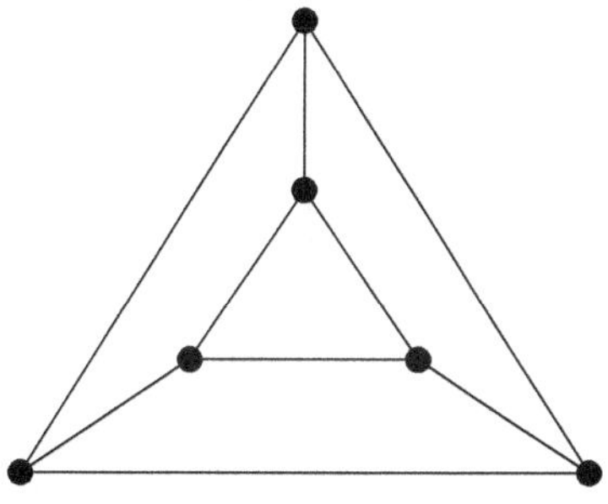

Figure 1.14

Remark 1.9. *A complete graph K_n is $n-1$ regular. We observe that a complete graph is regular*

9. **Bipartite Graph**

A simple graph G is called bipartite if its vertex set V can be partitioned into two non-empty disjoint sets V_1 and V_2 ($V = V_1 \cup V_2$, $V_1 \cap V_2 = \phi$) such that every edge in the graph connects a vertex in V_1 and a vertex in V_2. The graphs shown below are bipartite graph.

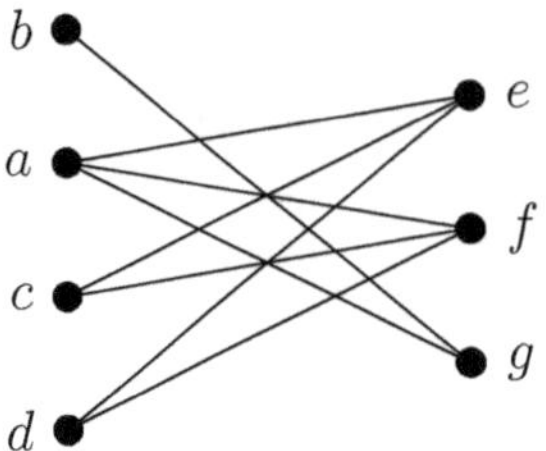

Figure 1.15

10. **Complete Bipartite Graph**

A bipartite graph in which there exists an edge between every vertex in V_1 and every vertex in V_2 is called a **complete bipartite graph.**

A complete bipartite graph with partitions V_1 and V_2 having m vertices in V_1 and n vertices in V_2 is denoted by $K_{m,n}$. The graph $K_{1,n}$ is called a star.

The complete bipartite graphs $K_{1,8}, K_{3,3}$ and $K_{3,5}$ are displayed in Figure 1.16.

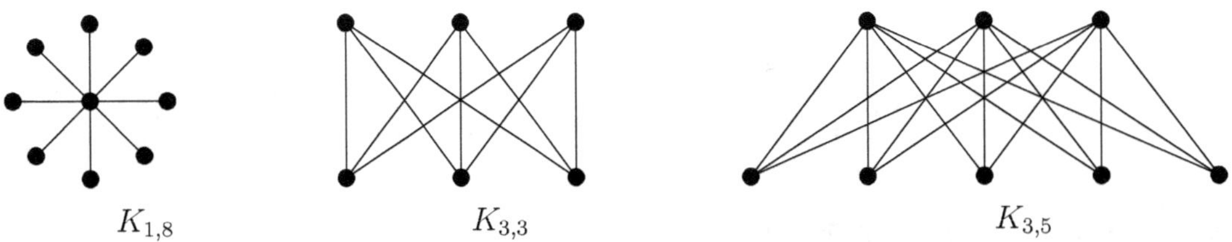

Figure 1.16

1.4 Types of Digraphs

1. **Simple digraph :** A digraph that has no self loop or parallel edges is called a simple digraph. Consider the digraph given below.

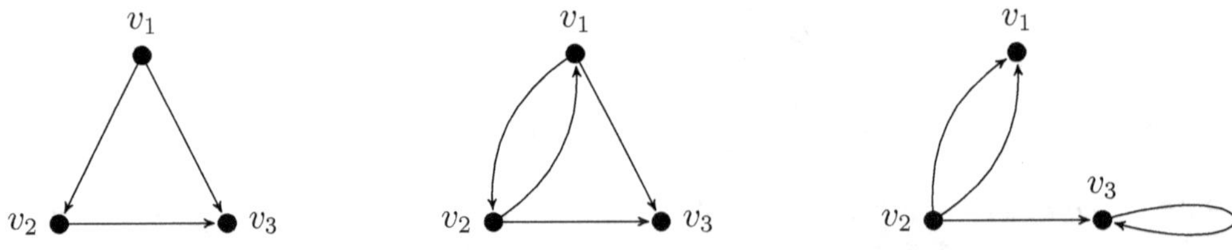

The digraphs G_1 and G_2 are simple digraphs, while G_3 is not simple since it contains a multiple edge as well as a loop.

2. **Symmetric digraph :** A digraph in which for every edge (a,b) there is also an edge (b,a) is called a symmetric digraph.

   The digraphs G_1 and G_2 given below are symmetric digraphs. But G_3 is not symmetric.

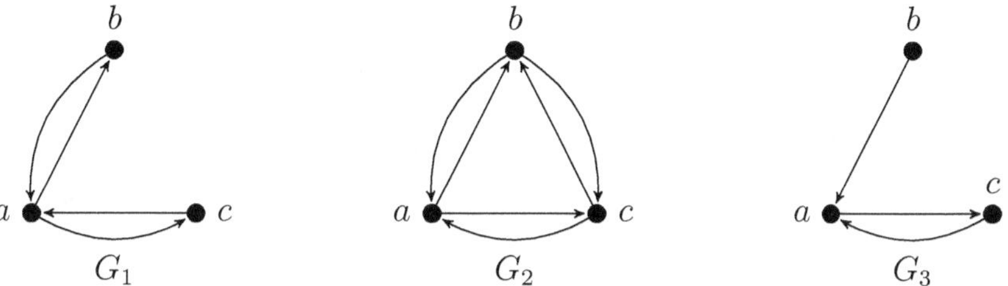

Figure 1.17

3. **Asymmetric digraph :** A digraph that has almost one directed edge between a pair of vertices with may or may not self loop is called a asymmetric digraph.

   The digraphs G_1 and G_2 given below are asymmetric digraphs.

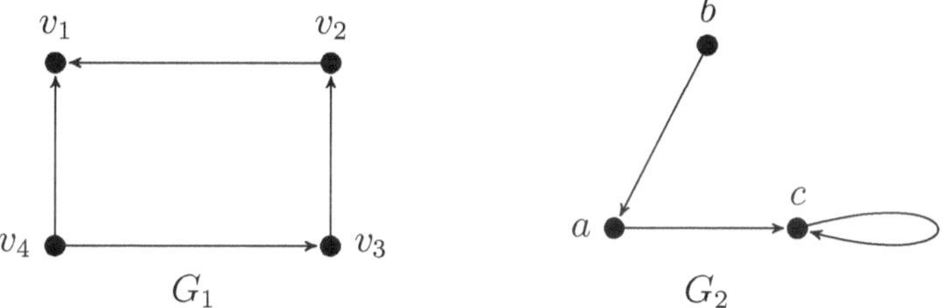

Figure 1.18

4. **Complete symmetric digraph :** A simple digraph in which there is exactly one directed edge from every vertex to every other vertex is called as complete symmetric digraph.

   Following digraphs G_1 and G_2 are complete symmetric digraphs.

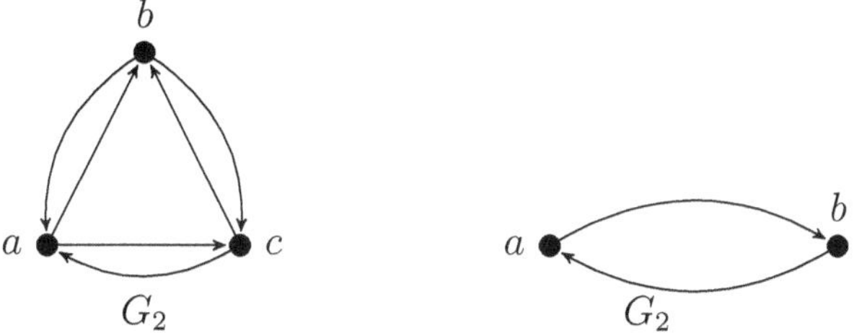

Figure 1.19

5. **Complete asymmetric digraph :** It is a digraph in which there is exactly one directed edge

The following digraph is a complete asymmetric digraph on 5 vertices.

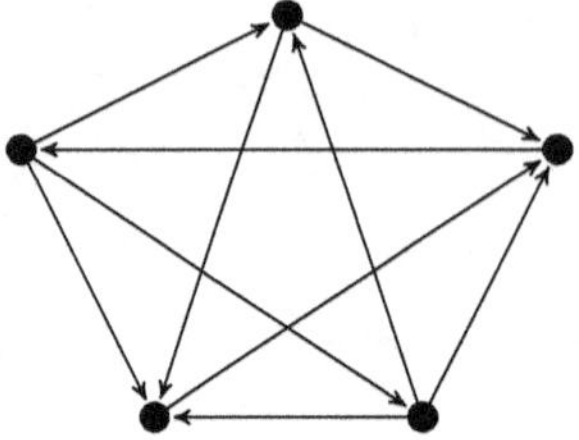

Figure 1.20

6. **Balanced digraph :** A digraph in which indegree and outdegree of each vertex is same is called as balanced digraph.

Consider the following graph.

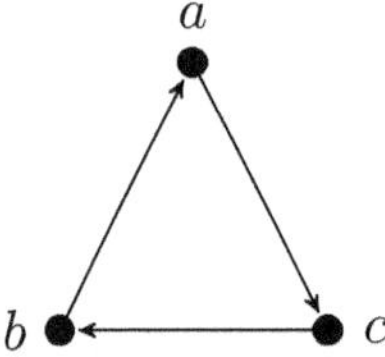

In this graph,

$$d^+(a) = d^-(a) = 1$$
$$d^+(b) = d^-(b) = 1$$
$$d^+(c) = d^-(c) = 1.$$

Hence it is a balanced digraph.

Definition 1.7. *The **degree sequence** of a graph is the sequence of the degrees of the vertices of the graph in nonincreasing order.*

For example, consider the following graph.

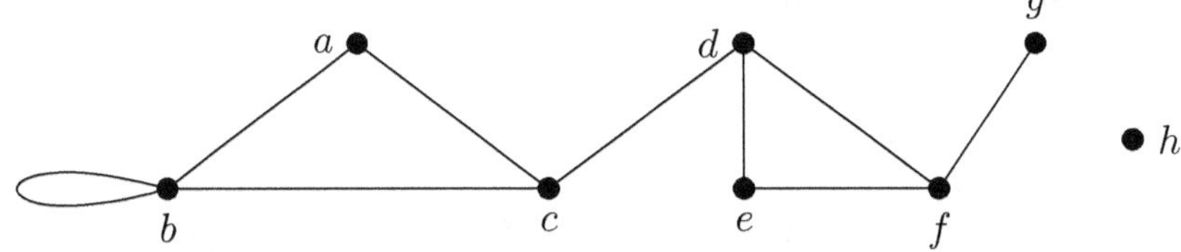

The degrees of the vertices are given as

$deg(a) = 2,\ deg(b) = 4,\ deg(c) = 3,\ deg(d) = 3,\ deg(e) = 2,\ deg(f) = 3,\ deg(g) = 1,\ deg(h) = 0.$

Therefore, the degree sequence of the graph is $4, 3, 3, 3, 2, 2, 1, 0$.

Definition 1.8. *A sequence $d_1, d_2, \cdots, d_n$ is called **graphical** if it is the degree sequence of a simple*

Steps to determine whether given degree sequence is graphical.

1. Consider the degree sequence in nonincreasing order.

2. If first term is $n \in \mathbb{N}$ then reduce next n terms by 1 counting from second term.

3. Rearrange the terms in nonincreasing order.

4. Continue steps 2 and 3.

5. If we get all terms zero, the original sequence is graphical otherwise it is not graphical.

Example 1.3. *Show that the sequence* $6, 6, 6, 6, 4, 3, 3, 0$ *is not graphical.*

Solution: We can reduce the sequence as follows :

Given sequence	$6, 6, 6, 6, 4, 3, 3, 0$
Reducing first 6 terms by 1 counting from second term	$5, 5, 5, 3, 2, 2, 0$
Reducing first 5 terms by 1 counting from second term	$4, 4, 2, 1, 1, 0$
Reducing first 4 terms by 1 counting from second term	$3, 1, 0, 0, 0.$

Since there exists no simple graph having one vertex of degree three and other vertex of degree one. The last sequence is not graphical. Hence the given sequence is also not graphical.

★ Illustrative Examples ★

Example 1.4. *Suppose G is a non-directed graph with 12 edges. If G has 6 vertices each of degree 3 and the rest have degree less than 3, what is the minimum number of vertices G can have?*

Solution: Number of edges in $G = 12$. Hence by handshaking lemma,

$$\sum deg(v_i) = 2|E| = 2 \times 12 = 24.$$

We have 6 vertices of degree 3. Let x denote the number of vertices each of whose degree is less than 3.

Then

$$\sum deg(v_i) < 6 \times 3 + 3x$$
$$\Rightarrow \qquad 24 < 18 + 3x$$
$$\Rightarrow \qquad 3x > 6$$
$$\Rightarrow \qquad x > 2.$$

The least positive integer for which the inequality $x > 2$ holds is $x = 3$.

Example 1.5. *Verify handshaking lemma for the graph shown in the following figure.*

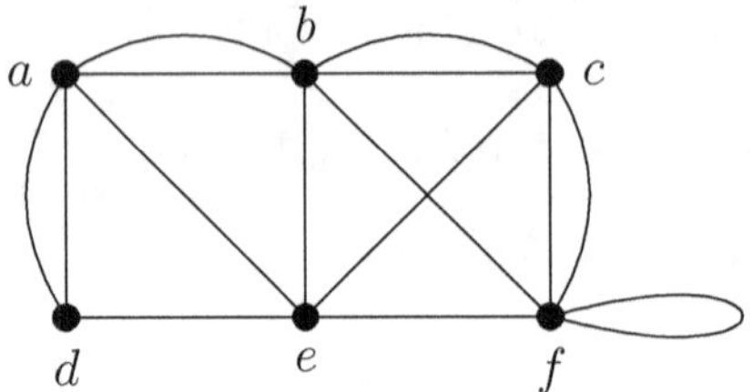

Figure 1.21

Solution: We have $V = \{a, b, c, d, e, f, g\}$ and total number of edges are 15, that is $|E| = 15$

$$\sum deg(v_i) = d(a) + d(b) + d(c) + d(d) + d(e) + d(f)$$
$$= 5 + 6 + 5 + 3 + 5 + 6$$
$$= 30$$
$$\Rightarrow \qquad \sum deg(v_i) = 30 = 2 \times 15 = 2|E|.$$

Hence handshaking lemma is verified.

Example 1.6. *Is it possible to construct a graph with* 12 *vertices such that* 2 *of the vertices have degree* 3 *and the remaining vertices have degree* 4.

Solution: Suppose it is possible to construct a graph with 12 vertices out of which 2 of them are having degree 3 and remaining vertices are having degree 4.

Hence by handshaking lemma,

$$\sum_{i=1}^{12} d(v_i) = 2e. \text{ where } e \text{ is the number of edges.}$$

According to given conditions

$$(2 \times 3) + (10 \times 4) = 2e$$
$$6 + 40 = 2e$$
$$2e = 46 \qquad\qquad \Rightarrow e = 23.$$

It is possible to construct a graph with 23 edges and 12 vertices which satisfy given conditions.

Example 1.7. *Is it possible to draw a simple graph with* 4 *vertices and* 7 *edges ? Justify.*

Solution: In a simple graph with n−vertices, the maximum number of edges will be $\dfrac{n(n-1)}{2}$.

Hence a simple graph with 4 vertices will have at most $\dfrac{4 \times 3}{2} = 6$ edges. Therefore, the simple graph with 4 vertices cannot have 7 edges.

Hence such a graph does not exist.

Example 1.8. *How many vertices and how many edges do the following graphs have ?*

(i) K_n (ii) C_n (iii) W_n (iv) $K_{m,n}$

Solution:

i) n vertices and $\dfrac{n(n-1)}{2}$ edges.

ii) n vertices and n edges.

iii) $n+1$ vertices and $2n$ edges.

iv) $m+n$ vertices and mn edges.

Example 1.9. *Does every regular graph complete? Justify.*

Solution: No. not every regular graph is complete. Consider following graph 2 - regular graph with 4 vertices which is not complete graph.

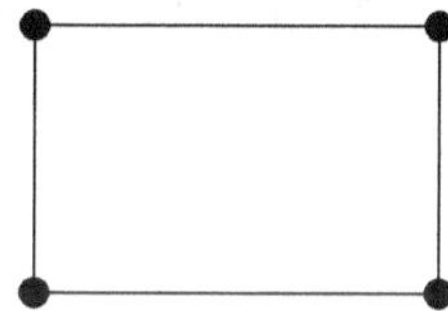

Example 1.10. *Determine which of the following graphs are bipartite.*

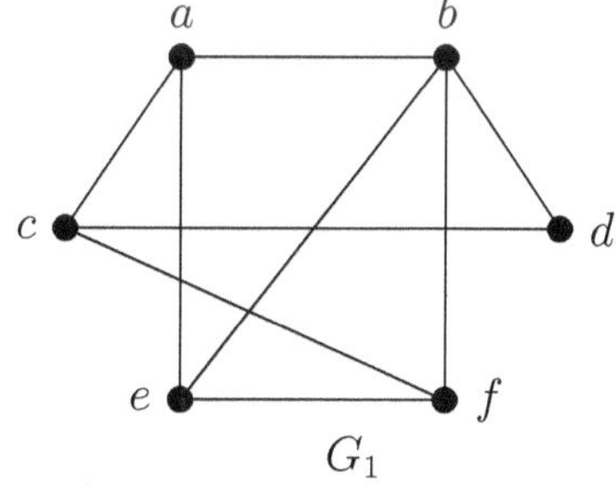

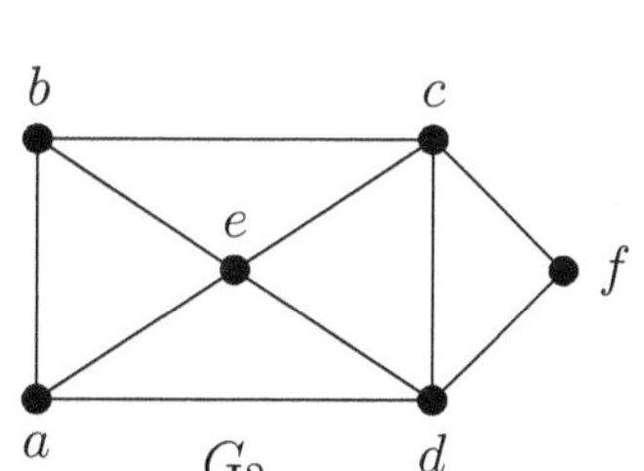

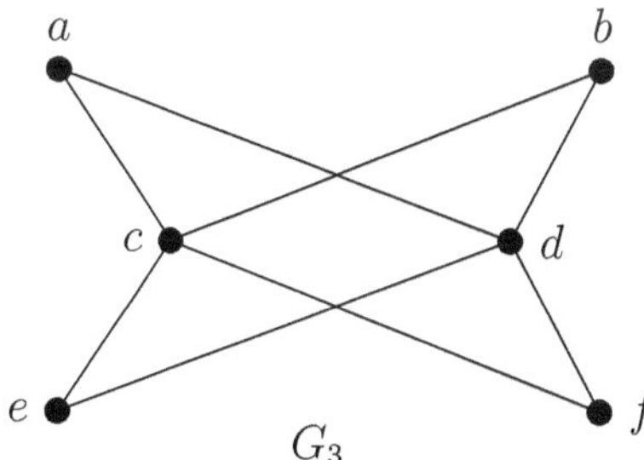

Figure 1.22

Solution:

i) A graph G_1 is not bipartite. We have $V = \{a, b, c, d, e, f\}$

To show that a graph is not bipartite we must give a proof that there is no possible way to specify the parts. We can show that this graph is not bipartite by the pigeonhole principle. Consider the vertices $b, e,$ and f. They form a triangle each is joined by an edge to the other two. By the pigeonhole principle, at least two of them must be in the same part of any proposed bipartition. Therefore there would be an edge joining two vertices in the same part, a contradiction to the

ii) The graph G_2 contains a triangle so by a similar argument as above it is not bipartite.

iii) G_3 is a bipartite graph with $V_1 = \{a, e, b, f\}, \quad V_2 = \{c, d\}$.

Example 1.11. *Determine whether each of these sequences is graphical. For those that are, draw a graph having the given degree sequence.*

$(i)\ 3, 3, 3, 3, 2$ $\qquad (ii)\ 6, 5, 4, 3, 2, 1$ $\qquad (iii)\ 6, 5, 5, 4, 3, 3, 2, 2, 2.$ $\qquad (iv)\ 2, 2, 2, 2, 2, 2$

Solution: i) We can reduce the sequence as follows :

Given sequence	$3, 3, 3, 3, 2$
Reducing first 3 terms by 1 counting from second term	$2, 2, 2, 2$
Reducing first 2 terms by 1 counting from second term	$1, 1, 2$
Rearranging terms we get	$2, 1, 1$
Reducing first 2 terms by 1 counting from second term	$0, 0.$

Since all terms are zero in the end, the given sequence is graphical. The graph corresponding to the sequence $3, 3, 3, 3, 2$ is given below

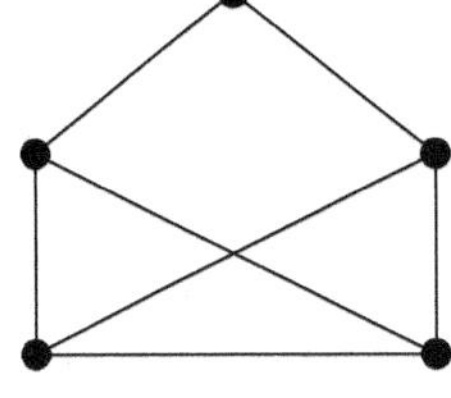

Figure 1.23

ii) We can reduce the sequence as follows :

Given sequence	$6, 5, 4, 3, 2, 1, 0$
Reducing first 6 terms by 1 counting from second term	$4, 3, 2, 1, 0, -1.$

Since this degree sequence contains one of the term as negative, the given sequence is not graphical.

iii) We can reduce the sequence as follows :

Given sequence	$6, 5, 5, 4, 3, 3, 2, 2, 2.$
Reducing first 6 terms by 1 counting from second term	$4, 4, 3, 2, 2, 1, 2, 2.$
Writing in decreasing order	$4, 4, 3, 2, 2, 2, 2, 1$
Reducing first 4 terms by 1 counting from second	$3, 2, 1, 1, 2, 2, 1$
Rearranging terms we get	$3, 2, 2, 2, 1, 1, 1$
Reducing first 3 terms by 1, counting from second	$1, 1, 1, 1, 1, 1.$

Sequence 1, 1, 1, 1, 1, 1 is graphical.

Hence the given sequence is also graphical.

The graph corresponding to the sequence $6, 5, 5, 4, 3, 3, 2, 2, 2$ is given below

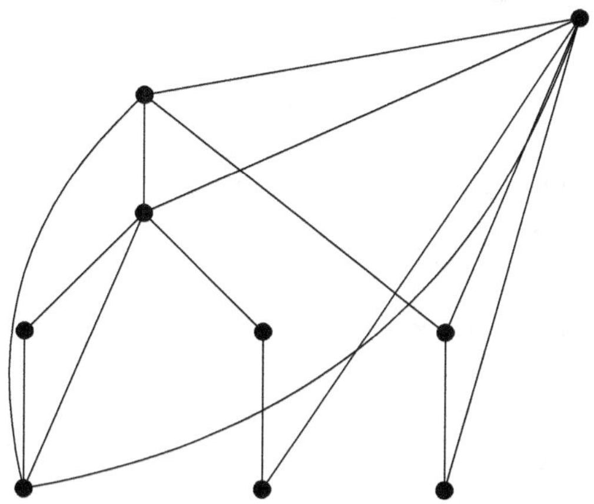

Figure 1.24

iv) We can reduce the sequence as follows :

Given sequence	$2, 2, 2, 2, 2, 2.$
Reducing first 2 terms by 1 counting from second term	$1, 1, 2, 2, 2.$
Writing in decreasing order	$2, 2, 2, 1, 1$
Reducing first 2 terms by 1 counting from second	$1, 1, 1, 1.$

Sequence $1, 1, 1, 1$ is graphical.

Hence the given sequence is also graphical.

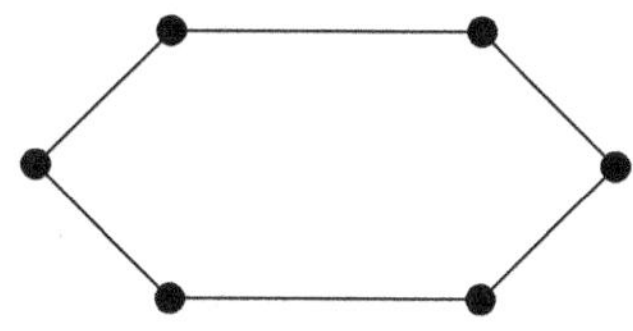

Figure 1.25

Example 1.12. *For which values of n are these graphs regular?*

(i) K_n (ii) C_n (iii) W_n (iv) Q_n

Solution:

i) The complete graph K_n is regular for all values of $n \geq 1$, since the degree of each vertex is $n - 1$.

ii) The degree of each vertex of C_n is 2 for all n for which C_n is defined, namely $n \geq 3$, so C_n is regular for all these values of n.

iii) The degree of the middle vertex of the wheel W_n is n, and the degree of the vertices on the

iv) The cube Q_n is regular for all values of $n \geq 0$, since the degree of each vertex in Q_n is n. (Note that Q_0 is the graph with 1 vertex.)

Exercise: 1.1

1. Give an example of the following or explain why no such example exists:

   i) a graph that has no odd vertices.

   ii) a noncomplete graph, all of whose vertices have degree 3.

2. Draw these graphs.

   $(i)\ K_7 \qquad (ii)\ K_{4,4} \qquad (iii)\ W_7 \qquad (v)\ C_7$

3. For which values of n are these graphs bipartite?

   $(i)\ K_n \qquad (ii)\ C_n \qquad (iii)\ W_n \qquad (iv)\ Q_n$

4. Draw a non-simple graphs G with degree sequence (1, 1, 3, 3, 3, 4, 6, 7).

5. A graph G has 21 edges, 3 vertices of degree 4 and other vertices are of degree 3. Find the number of vertices in G.

Solutions

1. i) Any cycle C_n $(n \geq 3)$
   ii)

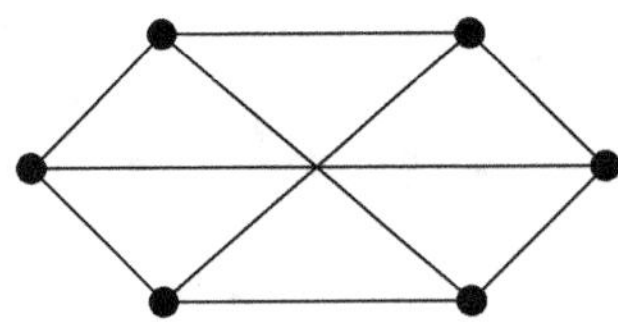

2.
    $i)$ $ii)$

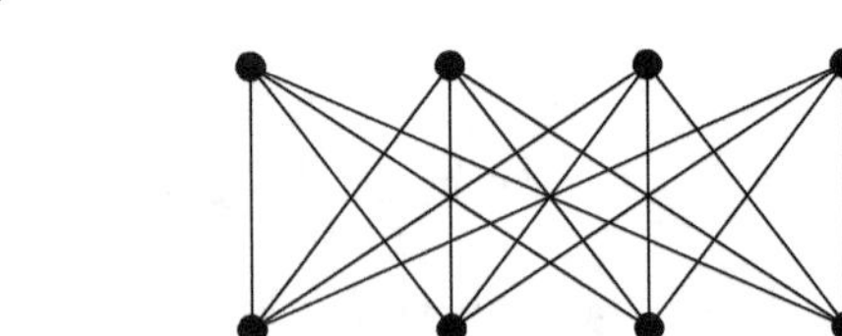

$K_7 \qquad\qquad\qquad\qquad\qquad K_{4,4}$

$$iii)\hspace{7cm}iv)$$

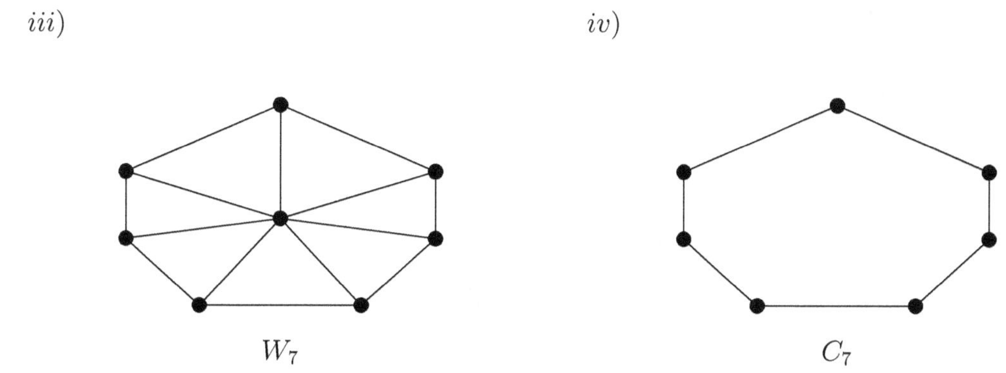

3. i) $n = 2$ ii) n is even and $n \geq 4$

   iii) For no n   iv) All n.

4.

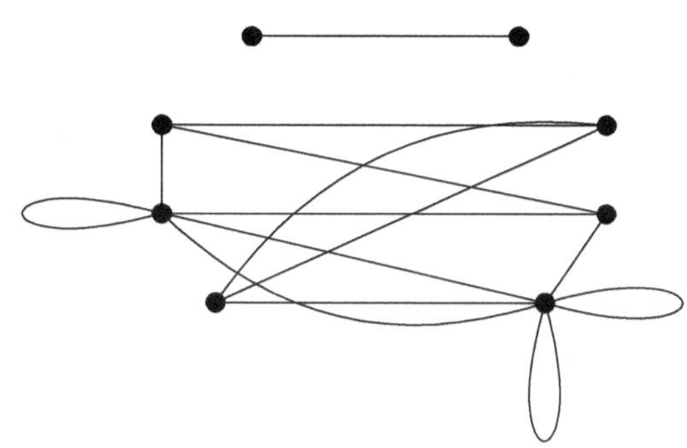

5. No of vertices $= 13$.

1.5 Representing Graphs and Graph Isomorphism

1.5.1 Representation of Graphs

There are many useful ways to represent graphs. In working with a graph it is helpful to be able to choose its most convenient representation. In this section we will show how to represent graphs in several different ways.

1. **Adjacency Lists**

Definition 1.9 (Adjacency). *In a graph G, if there is an edge e incident from the vertex u to the vertex v or incident on u and v, then the vertices u and v are said to be adjacent.*

In the graph of Following Fig.1.26 The vertices v_1 and v_2 are adjacent. Two non-parallel edges in a graph G, are said to be adjacent, if they are incident on the same vertex. In the graph of

Fig.1.26 The edges e_1 and e_2 are adjacent.

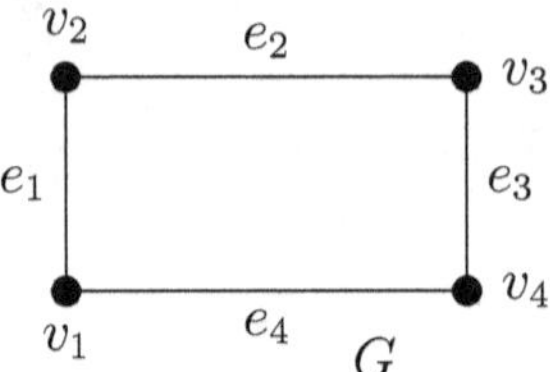

Figure 1.26

One way to represent a graph without multiple edges is to list all the edges of this graph. Another way to represent a graph with no multiple edges is to use adjacency lists, which specify the vertices that are adjacent to each vertex of the graph.

Example 1.13. *Use adjacency lists to describe the graphs given in Figure 1.27 .*

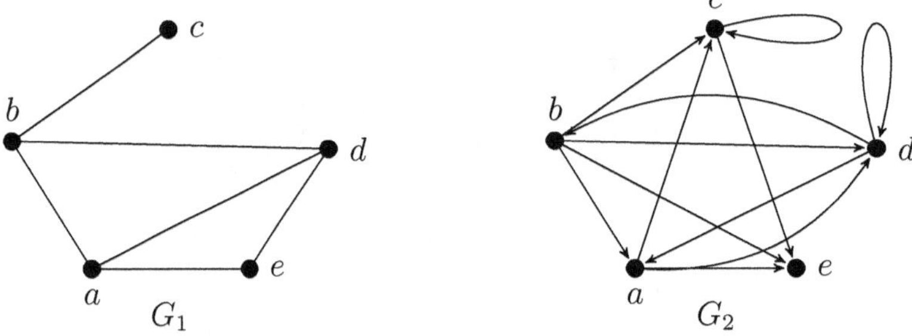

Figure 1.27

Solution:

Vertex	Adjacent Vertices
a	b,d,e
b	c, d,a
c	b
d	b,a,e
e	a,d

Table 1.1: Adjacency list of G_1

Vertex	Adjacent Vertices
a	c,d,e
b	c,d,a,e
c	c,e
d	d,b,a
e	

Table 1.2: Adjacency list of G_2

2. Adjacency Matrices

Two types of matrices commonly used to represent graphs will be presented here. One is based on the adjacency of vertices, and the other is based on incidence of vertices and edges.

Definition 1.10. *Let $G = (V, E)$ be a simple graph where $| V |= n$. Suppose that the vertices of G are listed arbitrarily as $v_1, v_2, \cdots, v_n$. The **adjacency matrix A (or $A(G)$)** of G, with respect to this listing of the vertices, is the $n \times n$ zeroone matrix with 1 as its (i, j)th entry when v_i and v_j are adjacent, and 0 as its (i, j)th entry when they are not adjacent. In other words, if its adjacency matrix is $A = [aij]$, then*

$$a_{ij} = \begin{cases} 1 & \text{if } \{v_i, v_j\} \text{ is an edge of } G, \\ 0 & \text{otherwise.} \end{cases}$$

Remark 1.10.   *1. For a given graph G, the adjacency matrix is based on ordering chosen for the vertices. Hence, there are as many as $n!$ different adjacency matrices for a graph with n vertices,*

2. Adjacency matrix A is symmetric i.e. $a_{ij} = a_{ji}$ for all i and j

3. The entries along the principal diagonal of A all zeros if and only if the graph has no self loops. A self loop at the vertex v_i is represented by 1 at $(i, i)^{th}$ entry.

Example 1.14. *Use adjacency matrix to represent the graphs shown in Figure 1.28 .*

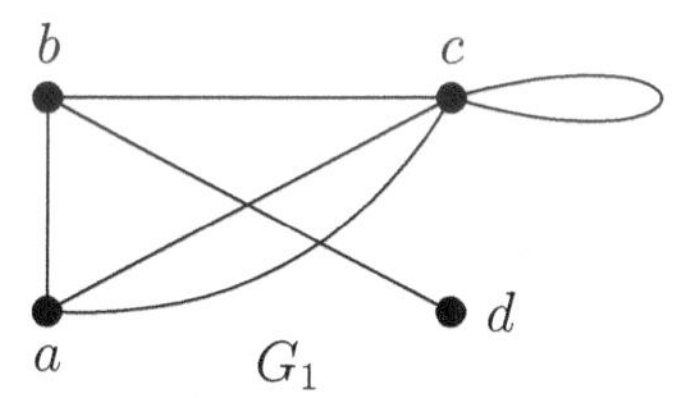
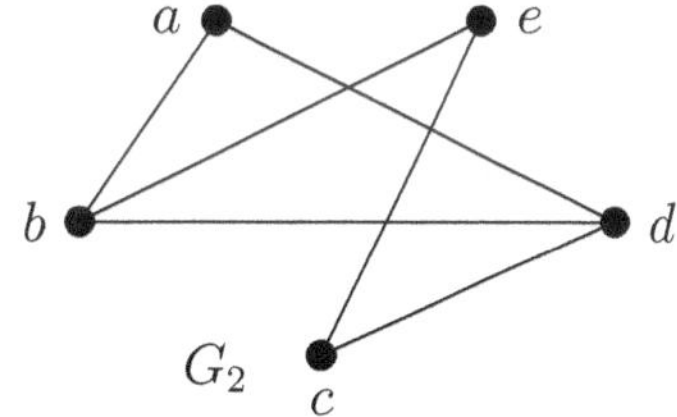

Figure 1.28

Solution:

$$A(G_1) = \begin{array}{c} \\ a \\ b \\ c \\ d \end{array} \begin{array}{c} \begin{array}{cccc} a & b & c & d \end{array} \\ \begin{bmatrix} 0 & 1 & 2 & 0 \\ 1 & 0 & 1 & 1 \\ 2 & 1 & 1 & 0 \\ 0 & 1 & 0 & 0 \end{bmatrix} \end{array}$$

$$A(G_2) = \begin{array}{c} \\ a \\ b \\ c \\ d \\ e \end{array} \begin{array}{c} \begin{array}{ccccc} a & b & c & d & e \end{array} \\ \begin{bmatrix} 0 & 1 & 0 & 1 & 0 \\ 1 & 0 & 0 & 1 & 1 \\ 0 & 0 & 0 & 1 & 1 \\ 1 & 1 & 1 & 0 & 0 \\ 0 & 1 & 1 & 0 & 0 \end{bmatrix} \end{array}$$

Definition 1.11 (Adjacency Matrix of a Directed graph). *Let $D = (V, E)$ be a digraph with n vertices, containing non-parallel edges. The adjacency matrix $A(G)$ of the digraph D is an $n \times n$ matrix defined by adjacency matrix is $A = [a_{ij}]$, then*

$$a_{ij} = \begin{cases} 1 & \text{if there is an edge directed from } v_i \text{ to } v_j \\ \end{cases}$$

Example 1.15. *Use adjacency matrix to represent the graphs shown in following figure.*

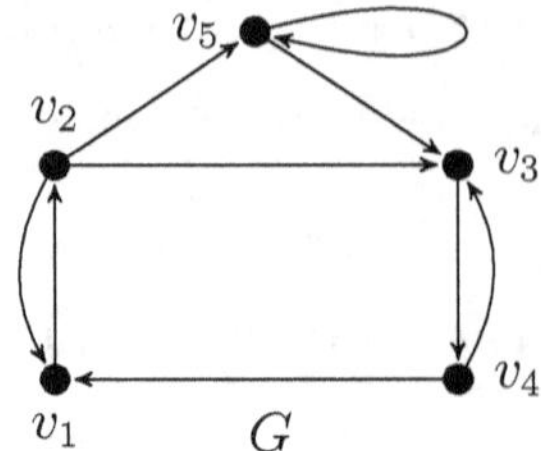

Solution:

$$A(G) = \begin{array}{c} \\ v_1 \\ v_2 \\ v_3 \\ v_4 \\ v_5 \end{array} \begin{array}{ccccc} v_1 & v_2 & v_3 & v_4 & v_5 \\ \begin{bmatrix} 0 & 1 & 0 & 0 & 0 \\ 1 & 0 & 1 & 0 & 1 \\ 0 & 0 & 0 & 1 & 0 \\ 1 & 0 & 1 & 0 & 0 \\ 0 & 0 & 1 & 0 & 1 \end{bmatrix} \end{array}$$

Definition 1.12 (Incidence Matrices). *Let $G = (V, E)$ be an undirected graph. Suppose that $v_1, v_2, \cdots, v_n$ are the vertices and $e_1, e_2, \cdots, e_m$ are the edges of G. Then the incidence matrix with respect to this ordering of V and E is the $n \times m$ matrix $M = [m_{ij}]$, where*

$$m_{ij} = \begin{cases} 1 & \text{when } e_j \text{ is incident with } v_i, \\ 0 & \text{otherwise.} \end{cases}$$

Example 1.16. *Use incidence matrix to represent the graphs shown in Figure 1.29 .*

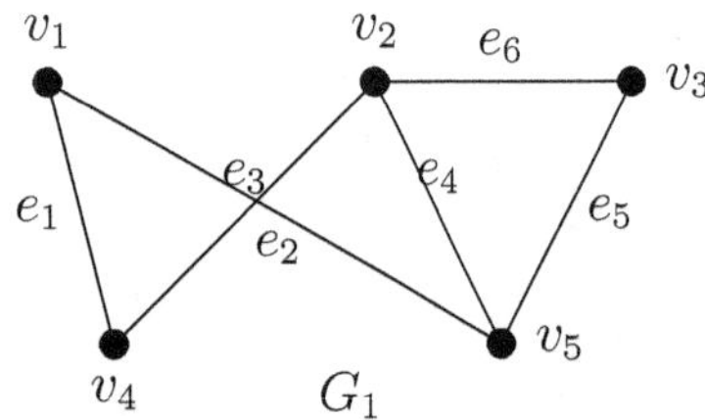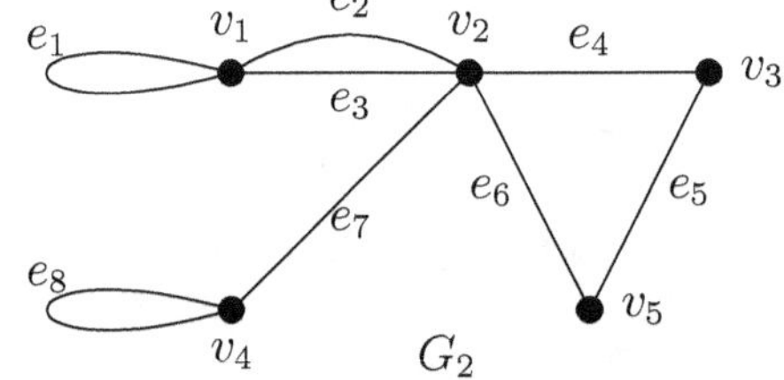

Figure 1.29

Solution:

$$I(G_1) = \begin{array}{c} \\ v_1 \\ v_2 \\ v_3 \\ v_4 \\ v_5 \end{array} \begin{array}{cccccc} e_1 & e_2 & e_3 & e_4 & e_5 & e_6 \\ \begin{bmatrix} 1 & 1 & 0 & 0 & 0 & 0 \\ 0 & 0 & 1 & 1 & 0 & 1 \\ 0 & 0 & 0 & 0 & 1 & 1 \\ 1 & 0 & 1 & 0 & 0 & 0 \\ 0 & 1 & 0 & 1 & 1 & 0 \end{bmatrix} \end{array}$$

$$I(G_2) = \begin{array}{c} \\ v_1 \\ v_2 \\ v_3 \\ v_4 \\ v_5 \end{array} \begin{array}{cccccccc} e_1 & e_2 & e_3 & e_4 & e_5 & e_6 & e_7 & e_8 \\ \begin{bmatrix} 1 & 1 & 1 & 0 & 0 & 0 & 0 & 0 \\ 0 & 1 & 1 & 1 & 0 & 1 & 1 & 0 \\ 0 & 0 & 0 & 1 & 1 & 0 & 0 & 0 \\ 0 & 0 & 0 & 0 & 0 & 0 & 1 & 1 \\ 0 & 0 & 0 & 0 & 1 & 1 & 0 & 0 \end{bmatrix} \end{array}$$

1.5.2 Isomorphism of Graphs

Definition 1.13. *Two graphs $G_1(V_1, E_1)$ and $G_2(V_2, E_2)$ are said to be **isomorphic** if following*

1. There exists a one to one and onto function $f : V_1 \to V_2$

2. There exists a one to one and onto function $g : E_1 \to E_2$ such that

3. e is an edge between u and v in G_1 if and only if $g(e)$ is an edge between $f(u)$ and $f(v)$.

We say that G_1 is isomorphic to G_2 and we denote it by $G_1 \cong G_2$.

Example 1.17. *Show that the following two graphs are isomorphic.*

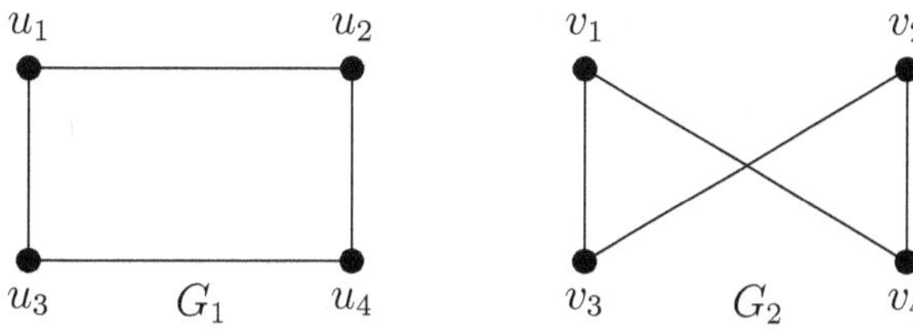

Figure 1.30

Solution: First we label edges in both graphs to establish the bijective mapping between the edge sets in such a way that the incidence relationship is preserved.

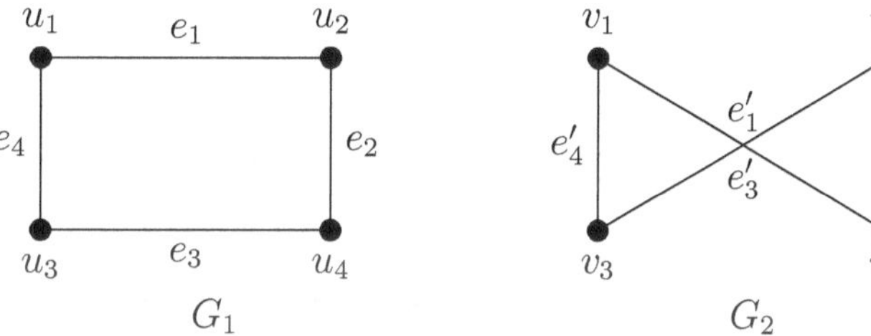

We define mapping between vertex set and edge set of G_1 and G_2 as follows

$$f : V_1 \to V_2 \qquad\qquad g : E_1 \to E_2$$

$$
\begin{aligned}
f(u_1) &= v_1 & g(e_1) &= e'_1 \\
f(u_2) &= v_4 & g(e_2) &= e'_2 \\
f(u_3) &= v_3 & g(e_3) &= e'_3 \\
f(u_4) &= v_2 & g(e_4) &= e'_4.
\end{aligned}
$$

Hence $G_1 \cong G_2$.

Example 1.18. *Determine whether the graphs shown in Figure 1.31 are isomorphic.*

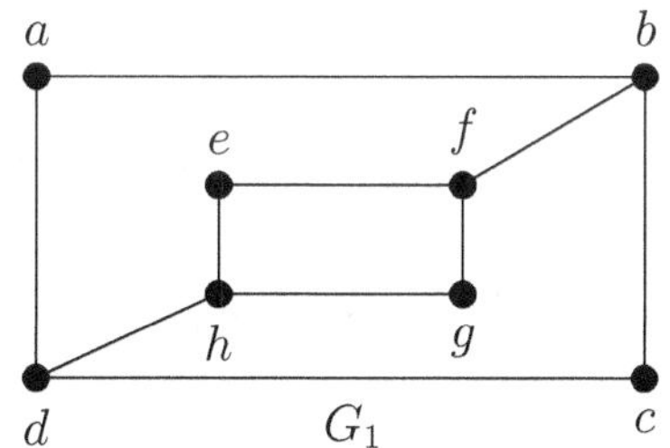
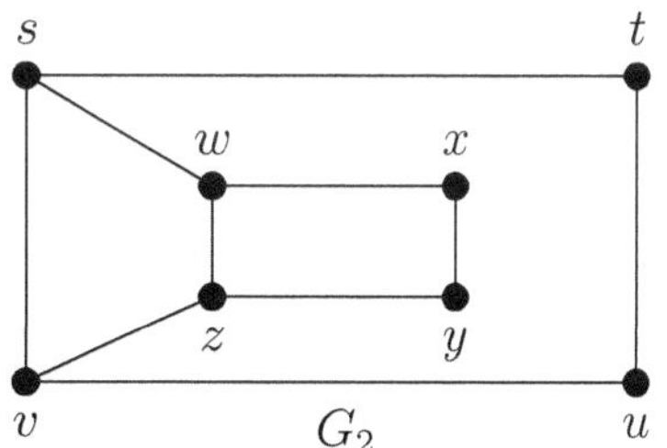

Solution: The graphs G_1 and G_2 both have 8 vertices and 10 edges. They also both have 4 vertices of degree 2 and 4 of degree 2. Because these invariants all agree, it is still conceivable that these graphs are isomorphic.

However, G_1 and G_2 are **not isomorphic.** To see this, in graph G_2 there are 4 vertices of deg 3 that form a circuit. But no such 4 vertices exists in G_1

★ <u>Illustrative Examples</u> ★

Example 1.19. *Draw a graph represented by adjacency matrix given below.*

$$A(G_1) = \begin{bmatrix} 0 & 1 & 1 & 1 \\ 1 & 0 & 1 & 0 \\ 1 & 1 & 0 & 1 \\ 1 & 0 & 1 & 0 \end{bmatrix} \qquad A(G_2) = \begin{bmatrix} 1 & 1 & 1 & 2 \\ 1 & 0 & 0 & 0 \\ 1 & 0 & 0 & 2 \\ 2 & 0 & 2 & 2 \end{bmatrix}$$

$$A(G_3) = \begin{bmatrix} 0 & 1 & 1 & 1 & 0 \\ 1 & 0 & 0 & 1 & 0 \\ 1 & 0 & 0 & 1 & 1 \\ 1 & 1 & 1 & 0 & 1 \\ 0 & 0 & 1 & 1 & 0 \end{bmatrix} \qquad A(G_4) = \begin{bmatrix} 1 & 1 & 1 & 2 & 0 \\ 1 & 1 & 2 & 1 & 1 \\ 1 & 2 & 1 & 1 & 1 \\ 2 & 1 & 1 & 1 & 0 \\ 0 & 1 & 1 & 0 & 1 \end{bmatrix}$$

Solution:

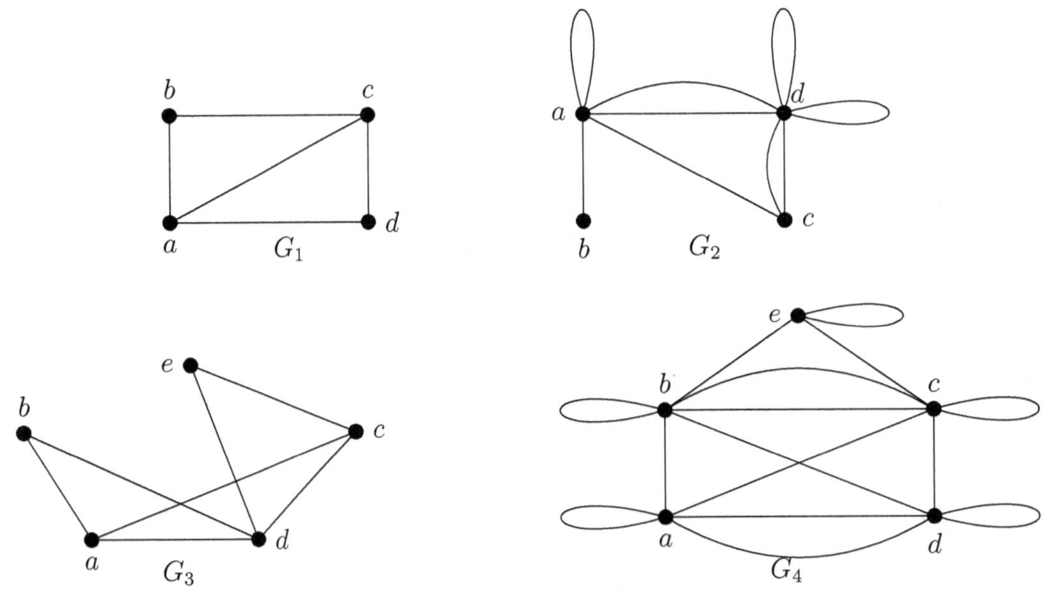

Example 1.20. *Use adjacency matrix to represent the graphs shown in following figure.*

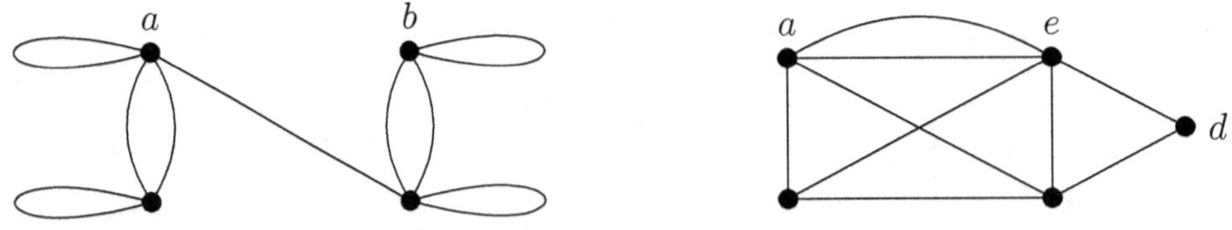

Solution:

$$A(G_1) = \begin{bmatrix} 1 & 0 & 2 & 1 \\ 0 & 1 & 0 & 2 \\ 2 & 0 & 1 & 0 \\ 1 & 2 & 0 & 1 \end{bmatrix} \qquad A(G_2) = \begin{bmatrix} 0 & 1 & 1 & 0 & 2 \\ 1 & 0 & 1 & 0 & 1 \\ 1 & 1 & 0 & 1 & 1 \\ 0 & 0 & 1 & 0 & 1 \\ 2 & 1 & 1 & 1 & 0 \end{bmatrix}$$

Example 1.21. *Find the adjacency matrix of the given directed graph shown in Figure 1.32 with respect to the vertices listed in alphabetic order. .*

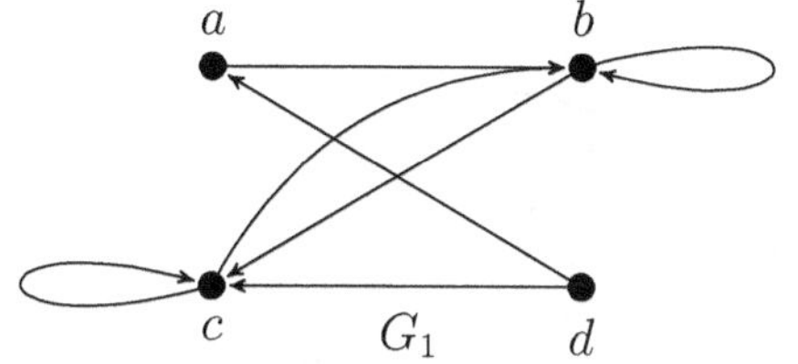

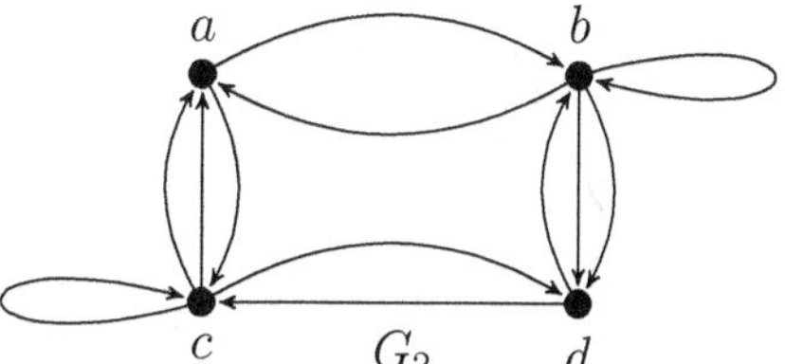

Figure 1.32

Solution:

$$A(G_1) = \begin{bmatrix} 0 & 1 & 0 & 0 \\ 0 & 1 & 1 & 0 \\ 0 & 1 & 1 & 0 \\ 1 & 0 & 1 & 0 \end{bmatrix} \qquad A(G_2) = \begin{bmatrix} 0 & 1 & 1 & 0 \\ 1 & 1 & 0 & 2 \\ 2 & 0 & 1 & 1 \\ 0 & 1 & 1 & 0 \end{bmatrix}$$

Example 1.22. *Use incidence matrix to represent the graphs shown below.*

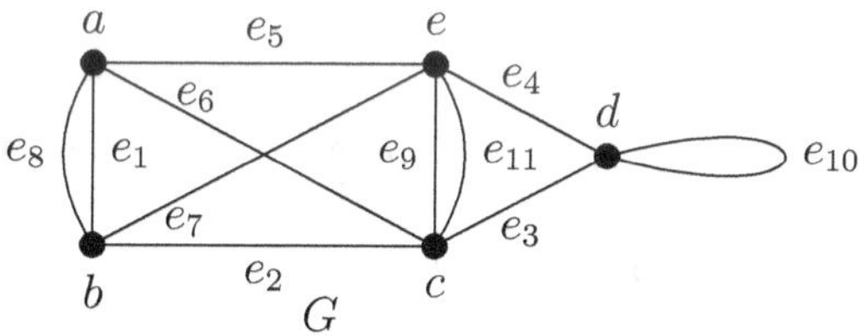

Figure 1.33

Solution:

$$I(G) = \begin{array}{c} \\ a \\ b \\ c \\ d \\ e \end{array} \begin{bmatrix} e_1 & e_2 & e_3 & e_4 & e_5 & e_6 & e_7 & e_8 & e_9 & e_{10} & e_{11} \\ 1 & 0 & 0 & 0 & 1 & 1 & 0 & 1 & 0 & 0 & 0 \\ 1 & 1 & 0 & 0 & 0 & 0 & 1 & 1 & 0 & 0 & 0 \\ 0 & 1 & 1 & 0 & 0 & 1 & 0 & 0 & 1 & 0 & 1 \\ 0 & 0 & 1 & 1 & 0 & 0 & 0 & 0 & 0 & 1 & 0 \\ 0 & 0 & 0 & 1 & 1 & 0 & 1 & 0 & 1 & 0 & 1 \end{bmatrix}$$

Example 1.23. *Draw the digraph of the following incidence matrices.*

$$i) \quad \begin{bmatrix} 0 & 1 & 0 & 1 & 0 \\ 1 & 0 & 0 & 0 & 0 \\ 0 & 1 & 0 & 0 & 0 \\ 1 & 0 & 1 & 0 & 1 \end{bmatrix} \qquad ii) \quad \begin{bmatrix} 0 & 0 & 0 & 1 & 0 & 1 & 0 & 0 \\ 0 & 0 & 0 & 0 & 1 & 1 & 1 & 1 \\ 0 & 0 & 0 & 0 & 0 & 0 & 0 & 1 \\ 1 & 1 & 1 & 0 & 1 & 0 & 0 & 0 \\ 0 & 0 & 1 & 1 & 0 & 0 & 1 & 0 \end{bmatrix}$$

Solution: First we label rows and columns of given matrices.

$$I(G_1) = \begin{array}{c} \\ v_1 \\ v_2 \\ v_3 \\ v_4 \\ v_5 \end{array} \begin{array}{ccccc} e_1 & e_2 & e_3 & e_4 & e_5 \\ \begin{bmatrix} 0 & 1 & 0 & 1 & 0 \\ 1 & 0 & 0 & 0 & 0 \\ 0 & 1 & 0 & 0 & 0 \\ 1 & 0 & 1 & 0 & 1 \\ 0 & 0 & 1 & 1 & 0 \end{bmatrix} \end{array}$$

$$I(G_2) = \begin{array}{c} \\ v_1 \\ v_2 \\ v_3 \\ v_4 \\ v_5 \\ v_6 \end{array} \begin{array}{cccccccc} e_1 & e_2 & e_3 & e_4 & e_5 & e_6 & e_7 & e_8 \\ \begin{bmatrix} 0 & 0 & 0 & 1 & 0 & 1 & 0 & 0 \\ 0 & 0 & 0 & 0 & 1 & 1 & 1 & 1 \\ 0 & 0 & 0 & 0 & 0 & 0 & 0 & 1 \\ 1 & 1 & 1 & 0 & 1 & 0 & 0 & 0 \\ 0 & 0 & 1 & 1 & 0 & 0 & 1 & 0 \\ 1 & 1 & 0 & 0 & 0 & 0 & 0 & 0 \end{bmatrix} \end{array}$$

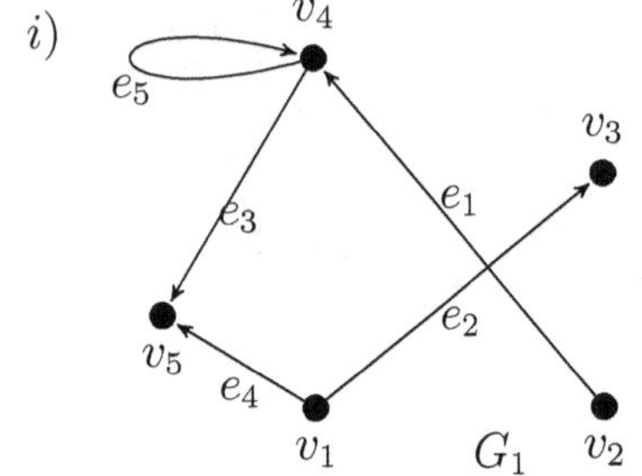
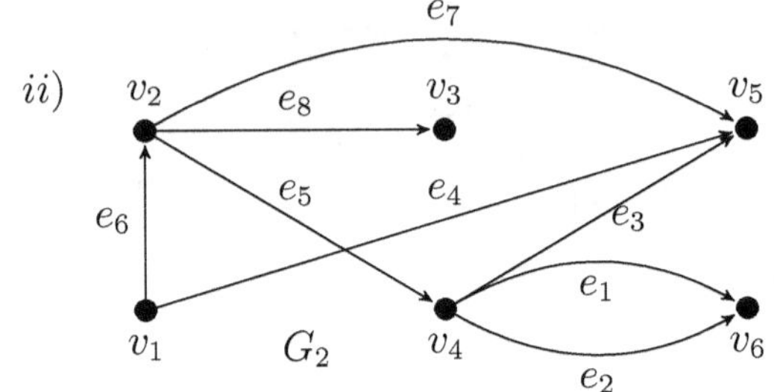

Figure 1.34

Example 1.24. *Determine whether the following pairs of graphs given in Figure 1.35 are isomorphic, If the graphs are not isomorphic give an invariant that the graphs do not share.*

a)

b)

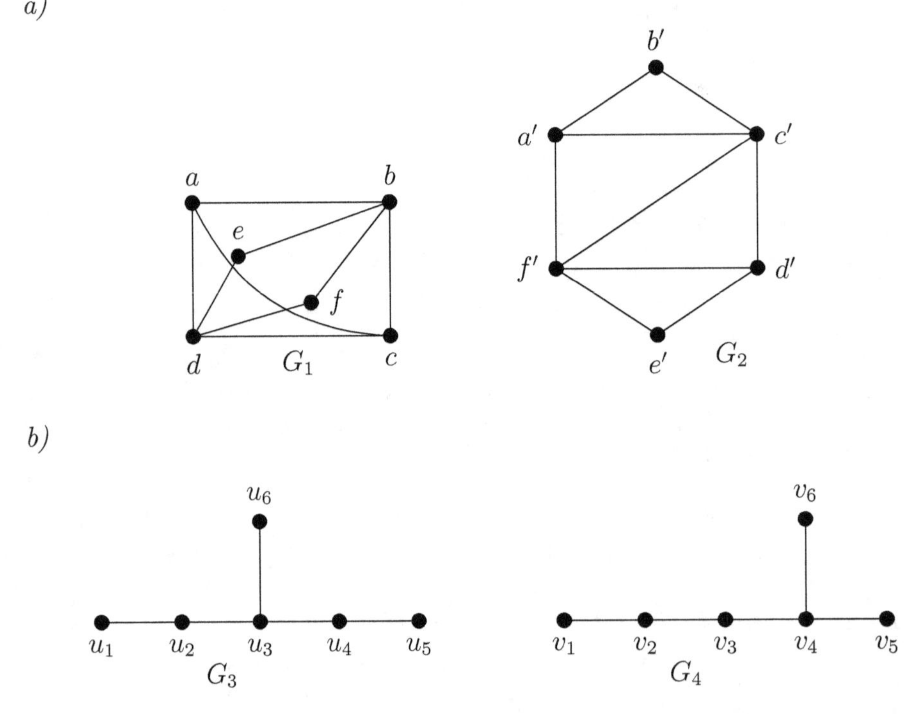

Figure 1.35

Solution:

a) Non isomorphic, In G_1 vertices of degree 3 are a and c and they are adjacent whereas in G_2 vertices of degree 3 are a' and d' are not adjacent.

b) Non isomorphic, In G_3 there is only one vertex of degree 3 namely u_3 which is incident with two vertices of degree 2 and one vertex of degree 1. But in G_4 only vertex of degree 3 (v_4) is incident with one vertex of degree 2 and two vertices of degree 1. Hence $G_3 \ncong G_4$.

Example 1.25. *Show that the following pair of graphs are isomorphic.*

a)

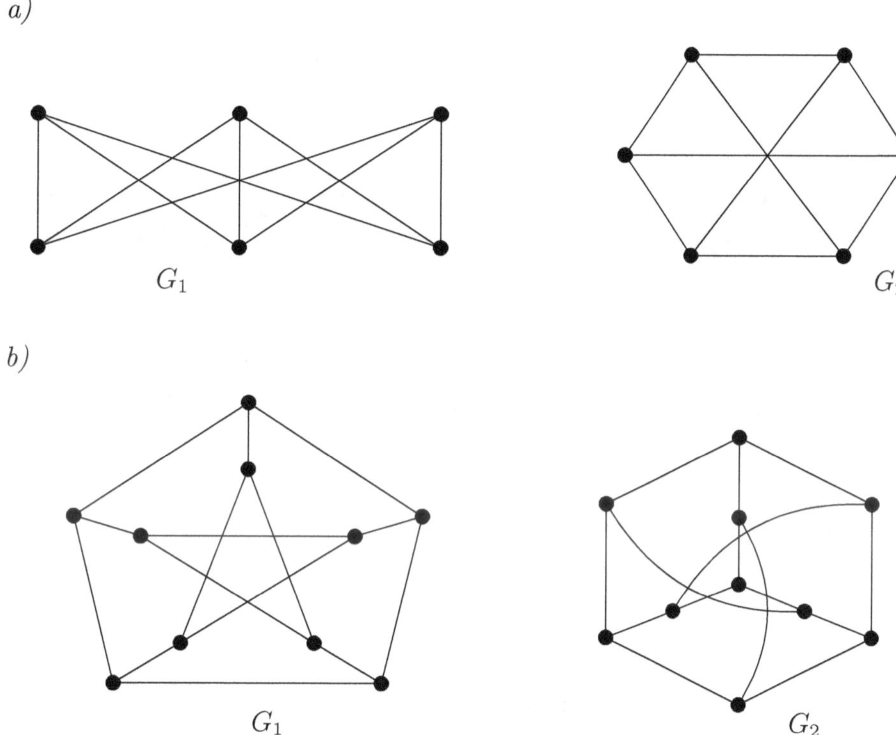

b)

Figure 1.36

Solution: First we label edges in both graphs to establish the bijective mapping between the edge sets in such a way that the incidence relationship is preserved.

a)

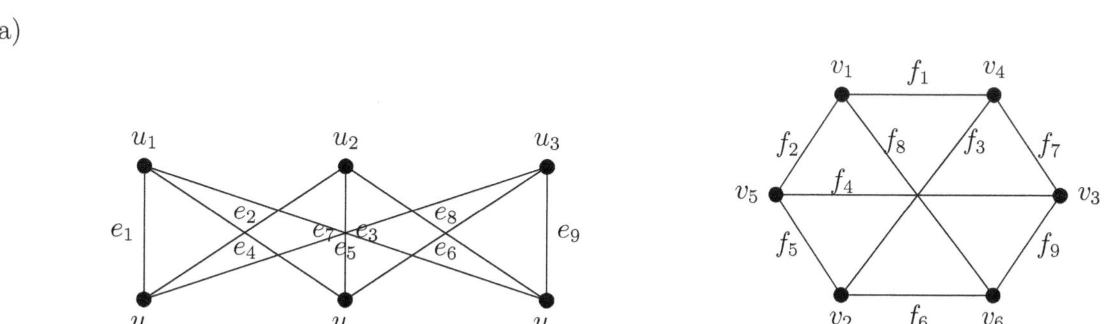

We define $f : V_1 \to V_2$ and $g : E_1 \to E_2$ by

$$f(u_i) = v_i \qquad\qquad i = 1, 2, 3, 4, 5, 6.$$

$$g(e_j) = f_j \qquad\qquad j = 1, 2, 3, \cdots, 9.$$

Hence $G_1 \cong G_2$.

b)

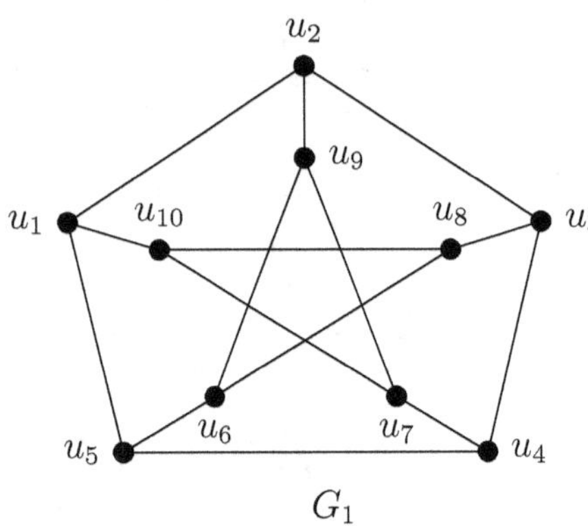

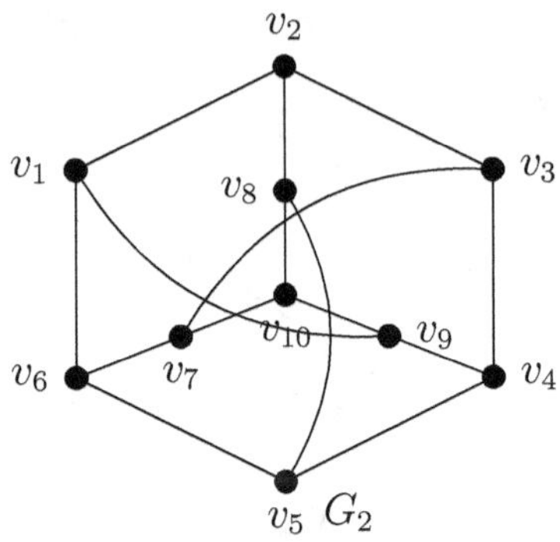

We check that the vertices adjacent to u_i in G_1 corresponds to the vertices adjacent to v_i in G_2. As both the graphs have no parallel edges, we don't label the edges in G_1 and G_2.

These are isomorphic graphs. One isomorphism is $f(u_1) = v_1, f(u_2) = v_9, f(u_3) = v_4, f(u_4) = v_3, f(u_5) = v_2, f(u_6) = v_8, f(u_7) = v_7, f(u_8) = v_5, f(u_9) = v_{10}$, and $f(u_{10}) = v_6$.

Exercise: 1.2

1. Use an adjacency list to represent the given graph.

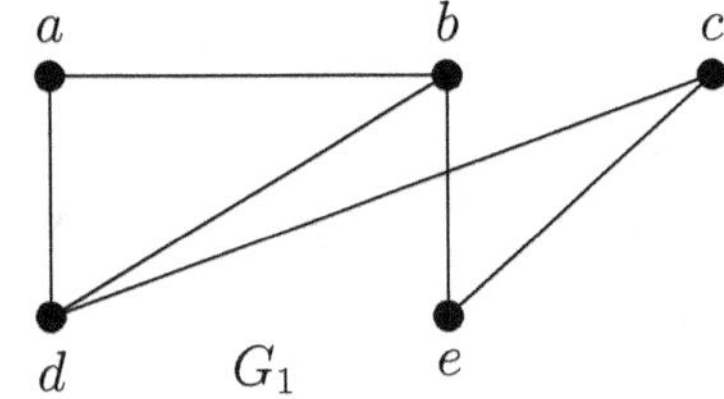

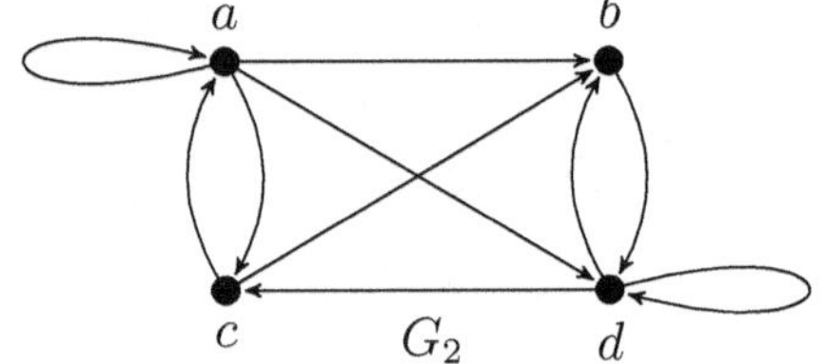

2. Represent each of these graphs with an adjacency matrix..

$$(i) \; K_4 \qquad (ii) \; K_{1,4} \qquad (iii) \; W_4 \qquad (iv) \; C_4.$$

3. Find adjacency matrix and incidence matrix of the following graphs.

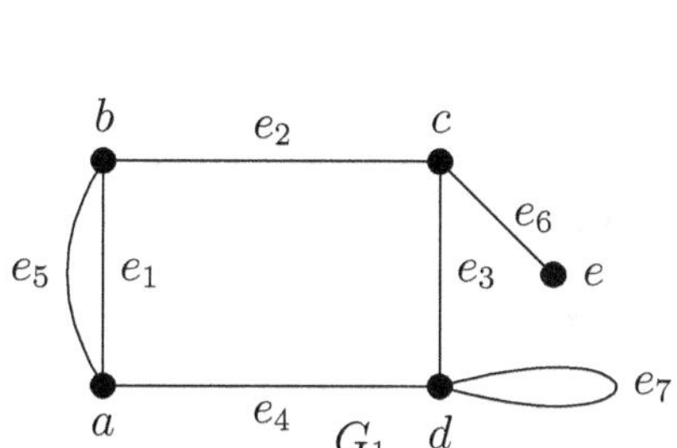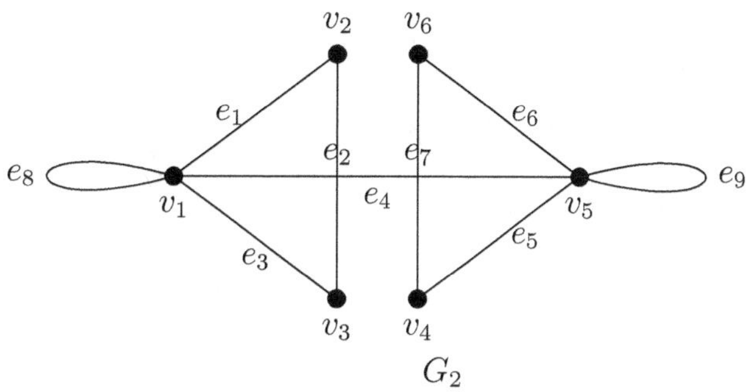

4. Find the adjacency matrix of the given directed graph.

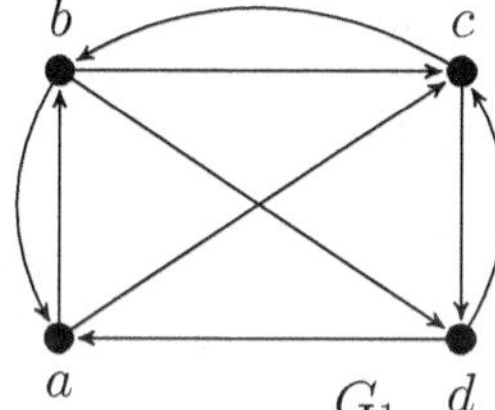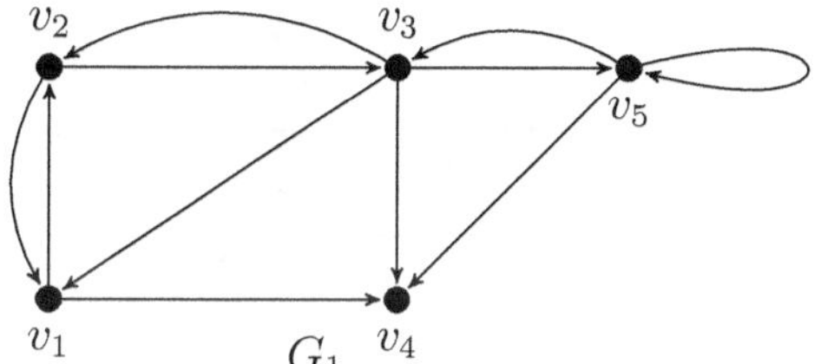

5. Draw an undirected graph represented by the given adjacency matrix.

$$i) \quad \begin{bmatrix} 1 & 3 & 2 \\ 3 & 0 & 3 \\ 2 & 3 & 0 \end{bmatrix} \qquad ii) \quad \begin{bmatrix} 1 & 2 & 0 & 1 \\ 2 & 0 & 3 & 0 \\ 0 & 3 & 1 & 1 \\ 1 & 0 & 1 & 0 \end{bmatrix}$$

6. Find the incidence matrix of the graph whose adjacency matrix is given as

$$A = \begin{array}{c} \\ a \\ b \\ c \\ d \\ e \\ f \end{array} \begin{array}{c} \begin{array}{cccccc} a & b & c & d & e & f \end{array} \\ \begin{bmatrix} 0 & 1 & 0 & 1 & 0 & 0 \\ 1 & 0 & 1 & 0 & 0 & 1 \\ 0 & 1 & 0 & 1 & 0 & 0 \\ 1 & 0 & 1 & 0 & 1 & 0 \\ 0 & 0 & 0 & 1 & 0 & 1 \\ 0 & 1 & 0 & 0 & 1 & 0 \end{bmatrix} \end{array}$$

7. Draw all simple non isomorphic graphs on 3 vertices.

8. Show that the following pair of graphs are isomorphic.

i)

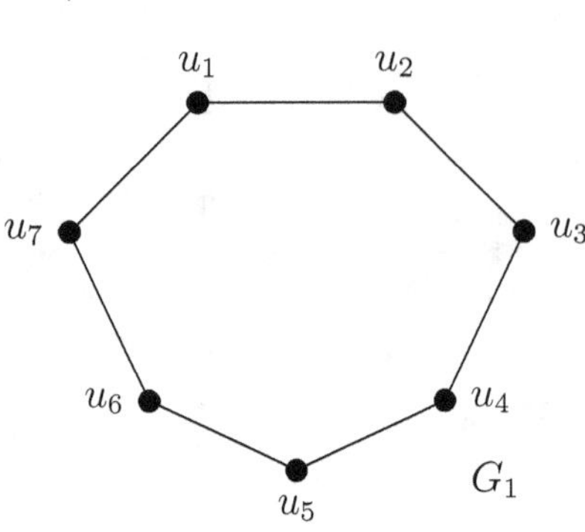

ii)

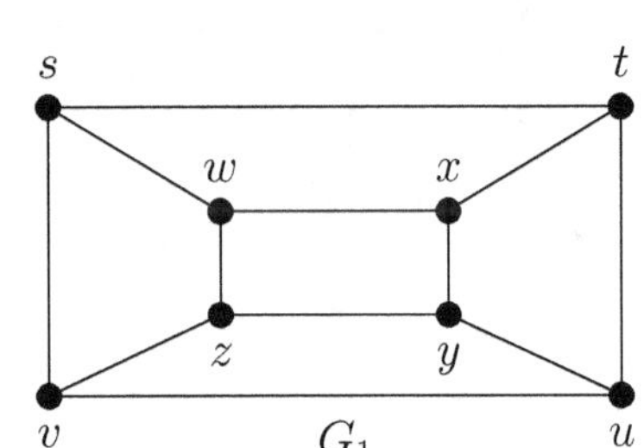

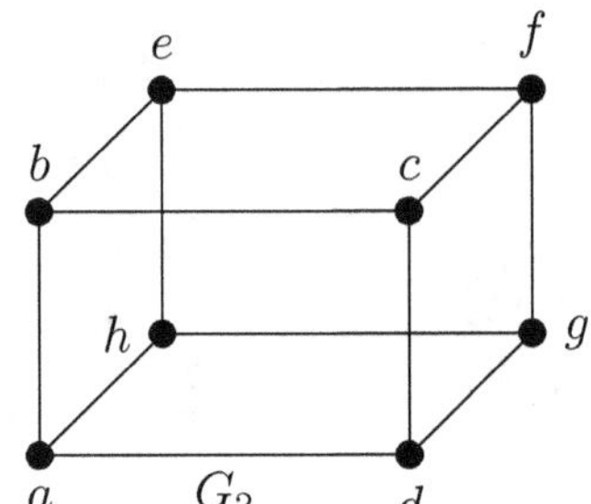

9. Determine the following pairs of graphs are isomorphic or not? Justify.

i)

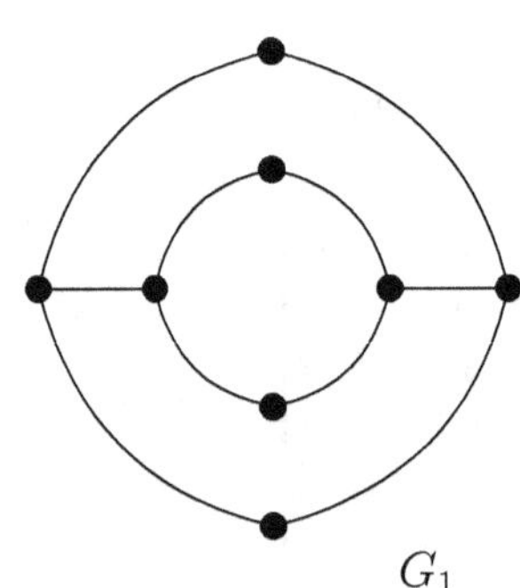

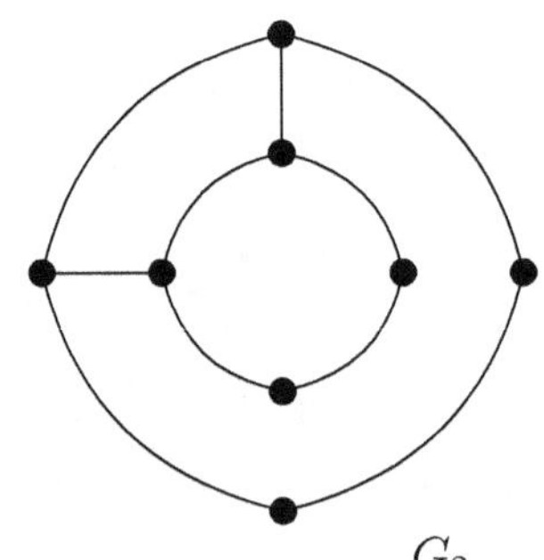

ii)

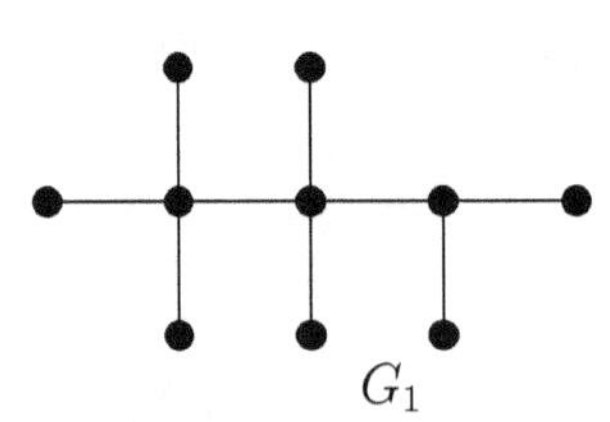

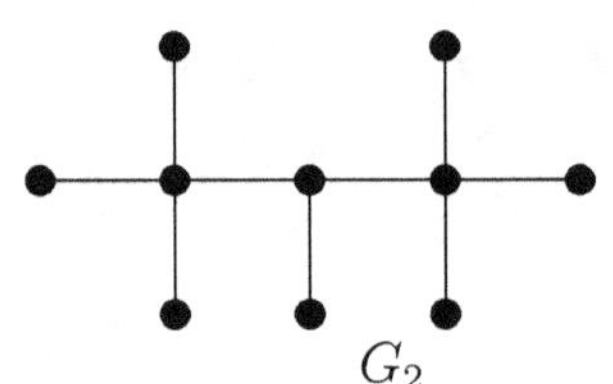

Solutions

3.

$$
A(G_1) = \begin{array}{c} \\ a \\ b \\ c \\ d \\ e \end{array}
\begin{array}{c} \begin{array}{ccccc} a & b & c & d & e \end{array} \\
\begin{bmatrix} 0 & 2 & 0 & 1 & 0 \\ 2 & 0 & 1 & 0 & 0 \\ 0 & 1 & 0 & 1 & 1 \\ 1 & 0 & 1 & 1 & 0 \\ 0 & 0 & 1 & 0 & 0 \end{bmatrix} \end{array}
\qquad
I(G_1) = \begin{array}{c} \\ a \\ b \\ c \\ d \\ e \end{array}
\begin{array}{c} \begin{array}{ccccccc} e_1 & e_2 & e_3 & e_4 & e_5 & e_6 & e_7 \end{array} \\
\begin{bmatrix} 1 & 0 & 0 & 1 & 1 & 0 & 0 \\ 1 & 1 & 0 & 0 & 1 & 0 & 0 \\ 0 & 1 & 1 & 0 & 0 & 1 & 0 \\ 0 & 0 & 1 & 1 & 0 & 0 & 1 \\ 0 & 0 & 0 & 0 & 0 & 1 & 0 \end{bmatrix} \end{array}
$$

$$
A(G_2) = \begin{array}{c} \\ v_1 \\ v_2 \\ v_3 \\ v_4 \\ v_5 \\ v_6 \end{array}
\begin{array}{c} \begin{array}{cccccc} v_1 & v_2 & v_3 & v_4 & v_5 & v_6 \end{array} \\
\begin{bmatrix} 1 & 1 & 1 & 0 & 1 & 0 \\ 1 & 0 & 1 & 0 & 0 & 0 \\ 1 & 1 & 0 & 0 & 0 & 0 \\ 0 & 0 & 0 & 0 & 1 & 1 \\ 1 & 0 & 0 & 1 & 1 & 1 \\ 0 & 0 & 0 & 1 & 1 & 0 \end{bmatrix} \end{array}
\qquad
I(G_2) = \begin{array}{c} \\ v_1 \\ v_2 \\ v_3 \\ v_4 \\ v_5 \\ v_6 \end{array}
\begin{array}{c} \begin{array}{ccccccccc} e_1 & e_2 & e_3 & e_4 & e_5 & e_6 & e_7 & e_8 & e_9 \end{array} \\
\begin{bmatrix} 1 & 0 & 1 & 1 & 0 & 0 & 0 & 1 & 0 \\ 1 & 1 & 0 & 0 & 0 & 0 & 0 & 0 & 0 \\ 0 & 1 & 1 & 0 & 0 & 0 & 0 & 0 & 0 \\ 0 & 0 & 0 & 0 & 1 & 0 & 1 & 0 & 0 \\ 0 & 0 & 0 & 1 & 1 & 1 & 0 & 0 & 1 \\ 0 & 0 & 0 & 0 & 0 & 1 & 1 & 0 & 0 \end{bmatrix} \end{array}
$$

4.

$$
A(G_1) = \begin{array}{c} \\ a \\ b \\ c \\ d \end{array}
\begin{array}{c} \begin{array}{cccc} a & b & c & d \end{array} \\
\begin{bmatrix} 0 & 1 & 1 & 0 \\ 1 & 0 & 1 & 1 \\ 0 & 1 & 0 & 1 \\ 1 & 0 & 1 & 0 \end{bmatrix} \end{array}
\qquad
A(G_2) = \begin{array}{c} \\ v_1 \\ v_2 \\ v_3 \\ v_4 \\ v_5 \end{array}
\begin{array}{c} \begin{array}{ccccc} v_1 & v_2 & v_3 & v_4 & v_5 \end{array} \\
\begin{bmatrix} 0 & 1 & 0 & 1 & 0 \\ 1 & 0 & 1 & 0 & 0 \\ 1 & 1 & 0 & 1 & 1 \\ 0 & 0 & 0 & 0 & 0 \\ 0 & 0 & 1 & 1 & 1 \end{bmatrix} \end{array}
$$

5.

$i)$

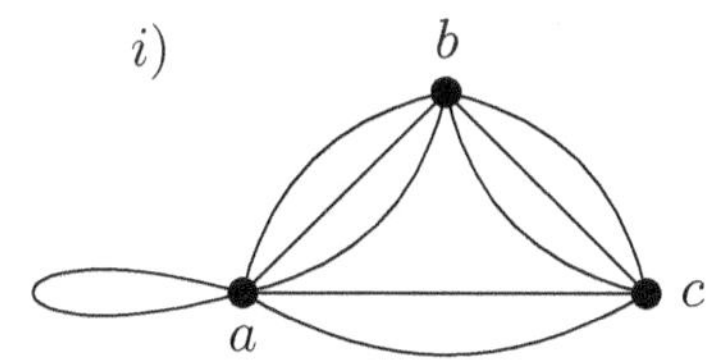

$ii)$

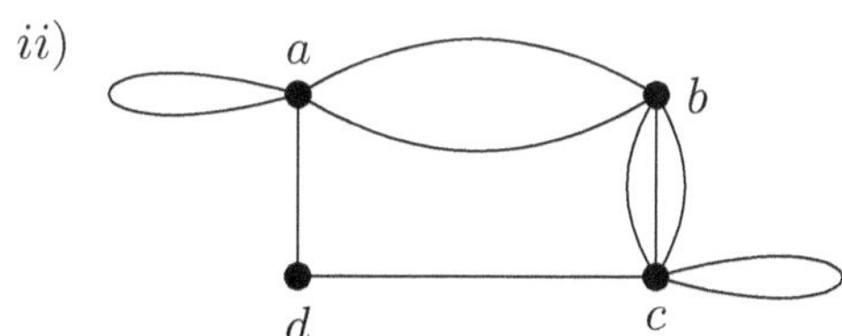

6. The corresponding graph is

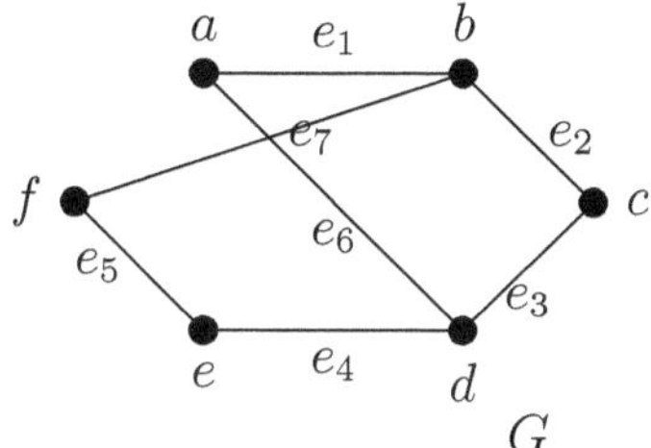

and required incidence matrix is

$$
I(G) = \begin{array}{c}
\quad\\a\\b\\c\\d\\e\\f
\end{array}
\begin{array}{ccccccc}
e_1 & e_2 & e_3 & e_4 & e_5 & e_6 & e_7\\
\left[\begin{array}{ccccccc}
1 & 0 & 0 & 0 & 0 & 1 & 0\\
1 & 1 & 0 & 0 & 0 & 0 & 1\\
0 & 1 & 1 & 0 & 0 & 0 & 0\\
0 & 0 & 1 & 1 & 0 & 1 & 0\\
0 & 0 & 0 & 1 & 1 & 0 & 0\\
0 & 0 & 0 & 0 & 1 & 0 & 1
\end{array}\right]
\end{array}
$$

7.  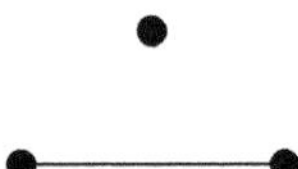 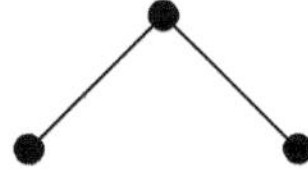 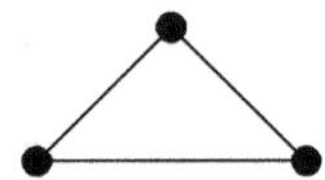

8. i)   These graphs are isomorphic, since each is the 7-cycle. One isomorphism is $f(u_1) = v_1$, $f(u_2) = v_3$, $f(u_3) = v_5$, $f(u_4) = v_7$, $f(u_5) = v_2$, $f(u_6) = v_4$, and $f(u_7) = v_6$.

ii) Graphs are isomorphic.

$f(a) = v$, $f(b) = s$, $f(c) = t$, $f(d) = u$, $f(h) = z$, $f(g) = y$, $f(f) = x$, and $f(e) = w$.

9. i)   In G_2 any two vertices of degree 3 are adjacent. This does not hold in G_1.

ii)   These graphs are not isomorphic, vertices of degree 4 in graph G_1 are adjacent whereas vertices of degree 4 are not adjacent in graph G_2.

1.6   Operations on Graphs

1.6.1   Subgraphs

Definition 1.14. *Let G and H be two graphs. The graph H is called a subgraph of G if $V(H)$ is a subset of $V(G)$ and $E(H)$ is a subset of $E(G)$.*

If H is a subgraph of G then

1. All the vertices of H are in G.

2. All the edges of H are in G.

3. Each edge of H has the same end points in H as in G.

Consider the graphs given below.

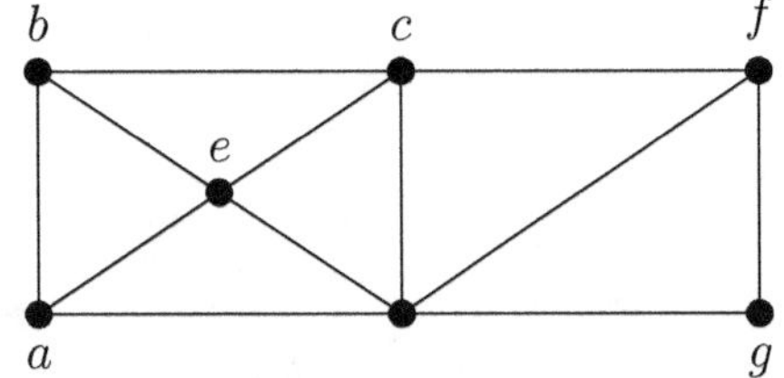 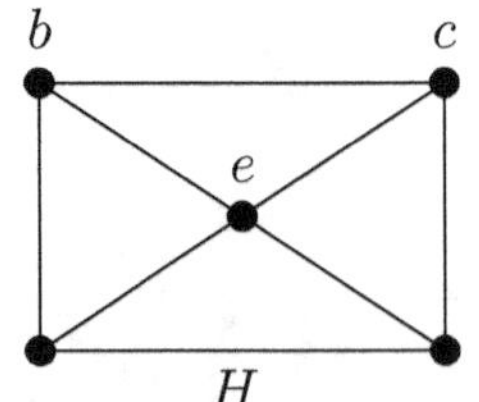 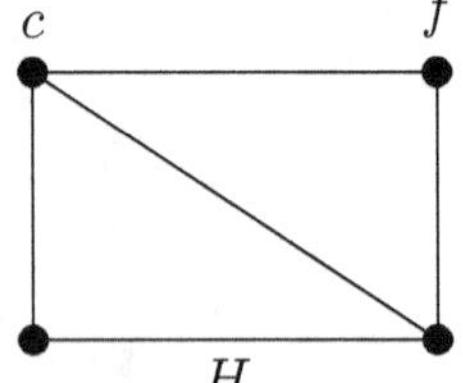

We observe that H_1 is a subgraph of G but H_2 is not a subgraph of G because edge joining c and g is in H_2 but not in G.

Definition 1.15 (Spanning Subgraph). *Let G and H be two graphs. A spanning subgraph H is a subgraph containing all the vertices of G.*

In other words, if $V(H) \subset V(G)$ and $E(H) \subseteq E(G)$ then H is a proper subgraph of G and if $V(H) = V(G)$ then we say that H is a spanning subgraph of G.

A spanning subgraph need not contain all the edges in G.

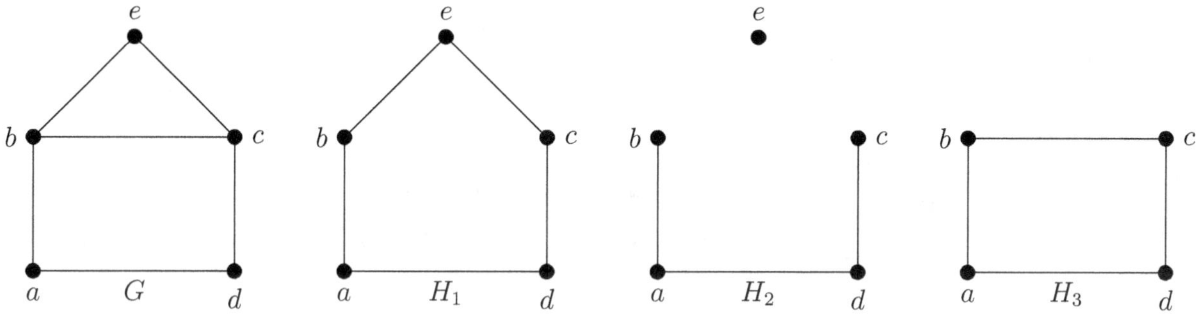

Figure 1.37

The graphs H_1 and H_2 of the above figure are spanning subgraphs of G, but H_3 is not a spanning subgraph of G.

Definition 1.16 (Edge Disjoint Subgraph). *Two subgraphs H_1 and H_2 of a graph G are called edge disjoint subgraphs if H_1 and H_2 do not share any edges in common.*

Definition 1.17 (Vertex Disjoint Subgraph). *If the subgraph H_1 and H_2 of G have no vetrex in common then they are called vertex disjoint subgraphs.*

Remark 1.11. *Two vertex disjoint subgraphs of a given graph G are also edge disjoint. Hence vertex disjoint subgraphs are called **disjoint subgraphs**.*

Consider the following graph

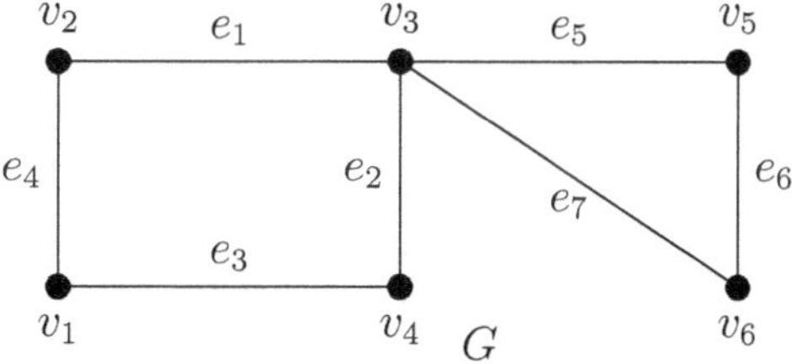

Figure 1.38

Consider the following subgraphs of G

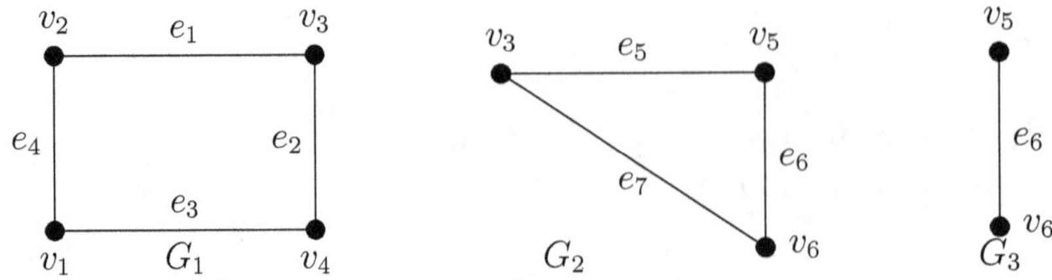

Figure 1.39

G_1 and G_2 are edge disjoint subgraphs but not vertex disjoint because v_2 is a common vertex.

G_1 and G_3 are vertex disjoint and hence disjoint subgraphs.

G_2 and G_3 are neither egde disjoint nor vertex disjoint.

1.6.2 Deletion of vertices and edges from a Graph

The removal of a vertex v_i from a graph G result in that subgraph $G - v_i$ of G containing of all vertices in G except v_i and all edges not incident with v_i. Thus $G - v_i$ is the maximal subgraph of G not containing v_i.

On the otherhand, the removal of an edge e_j from G yields the spanning subgraph $G - e_j$ containing all edges of G except e_j.Thus $G - e_j$ is the maximal subgraph of G not containing e_j.

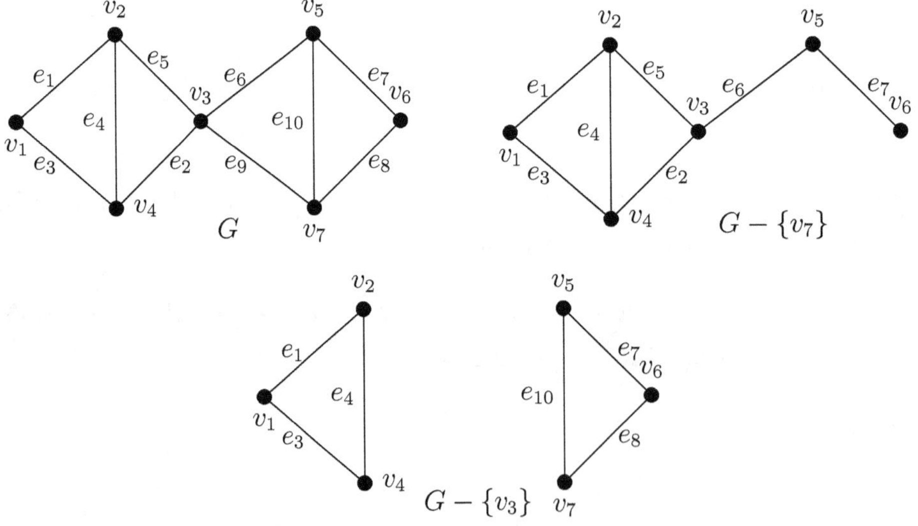

Figure 1.40: **Deleting vertices from a graph.**

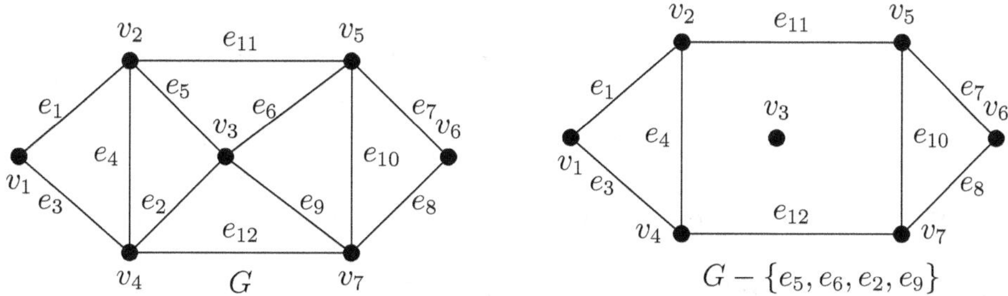

Figure 1.41: **Deleting edges from a graph.**

Definition 1.18. *Let $G(V, E)$ be a graph and $W = \{v_1, v_2, \cdots, v_i\} \subset V$. A subgraph of G, whose vertex set is W and edge set consists of those edges in G which has both end vertices in W is called a* **vertex induced subgraph.** *It is denoted by* $< W >$.

Consider the graph G given below having vertices $\{v_1, v_2, v_3, v_4, v_5\}$. Also let $W = \{v_2, v_3, v_4, v_1\}$.

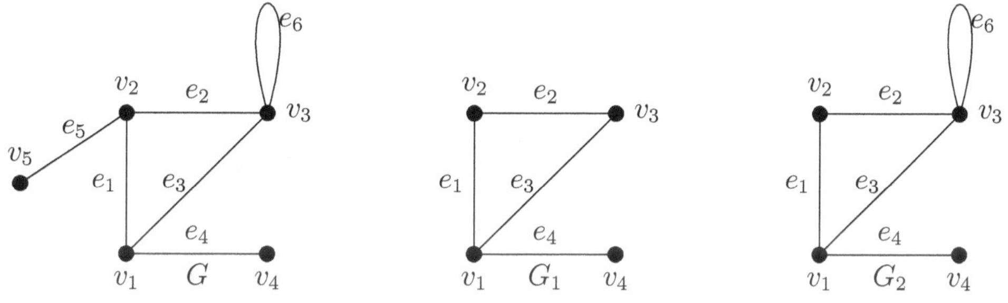

Figure 1.42

Here G_1 is not a vetrex induced subgraph since $e_6 \in E(G)$ but $e_6 \notin E(G_1)$. On the otherhand G_2 is a vertex induced subgraph of G, hence $G_2 = < W >$.

Definition 1.19. *Let $G(V, E)$ be a graph and $F = \{e_1, e_2, \cdots, e_j\} \subset E$. A subgraph of G, whose edge set is F and vertex set consists of those vertices in G which are end vertices of edges in F is called a* **edge induced subgraph.** *It is denoted by* $< F >$.

Consider the following graph G. Also let $F = \{e_1, e_2, e_4\}$.

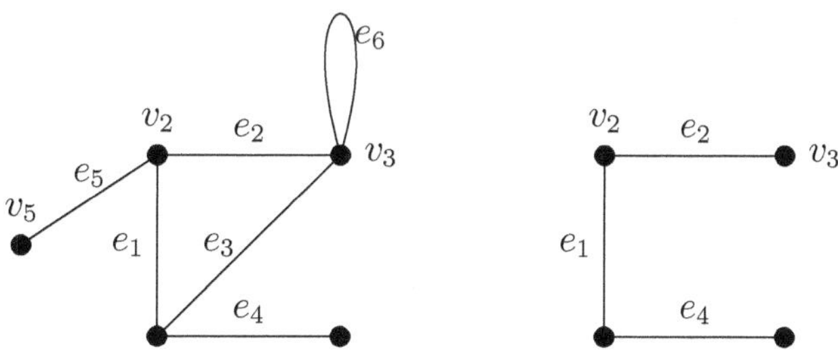

Here $G_1 = <F>$ is an edge induced subgraph of G.

Remark 1.12. *Let $G(V,E)$ be a graph and $|V| = m, |E| = n$. The total non-empty subsets of V*
is $2^m - 1$ and total subsets of E is 2^n.
Thus total number of subgraphs are equal to $(2^m - 1) \times 2^n$.
The number of spanning subgraphs are equal to 2^n.

1.6.3   Complement of a Graph

Definition 1.20. *The complement $\overline{G}$ of G is defined as a simple graph with the same vertex set as*
G where two vertices u and v adjacent in $\overline{G}$ only when they are not adjacent in G.

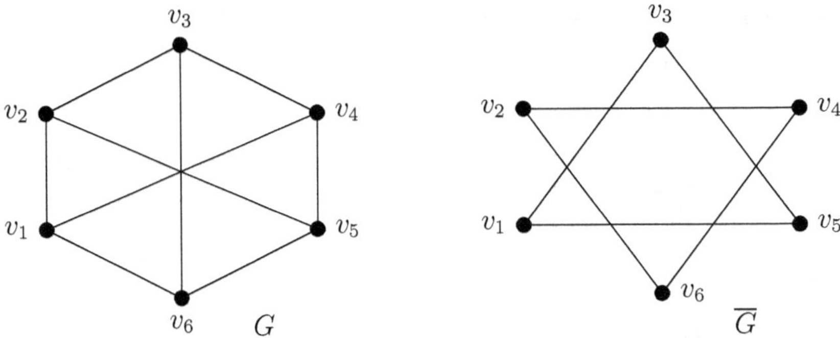

Figure 1.43

$\overline{G}$ is acomplement of graph G(V,E).

Remark 1.13.   *1. Complement of the complement of the graph G is G itself, i.e. $\overline{\overline{G}} = G$*

2. Complement of a complete graph is null graph and vice versa.

3. Complement of a regular graph is regular. If G is k-regular on n vertices, then $\overline{G}$ is $n - k - 1$
regular.

4. Let K_n be complete graph on n vertices and G be any graph on n vertices then
Number of edges in G + Number of edges in $\overline{G}$ = Number of edges in $K_n = \dfrac{n(n-1)}{2}$

Definition 1.21 (Self complementary graph).

If a simple graph G is isomorphic with its complement $\overline{G}$ (i.e.$G \cong \overline{G}$) then G is called a self comple-
mentary graph.

Consider a following graph G and its complement $\overline{G}$.

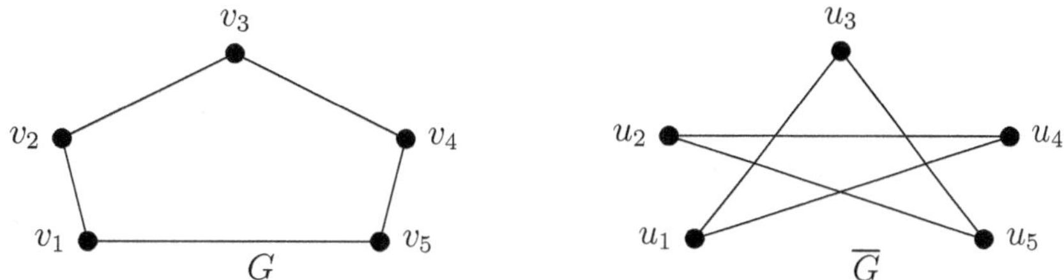

Figure 1.44

Here $G \cong \overline{G}$ as the bijection $v_1 \to u_1,\ v_2 \to u_2,\ v_3 \to u_3,\ v_4 \to u_4,\ v_5 \to u_5,$ between the vertices preserves adjacency, therefore G is self complementary graph.

Example 1.26. *Show that in a self complementary graph G on n vertices, n or $n-1$ must be a multiple of 4. (that is $n = 4k$ or $n = 4k+1$).*

Solution: Let G be a self complementary graph. Then $G \cong \overline{G}$, so both of G and $\overline{G}$ have the same number of edges. Also the total number of edges in G and $\overline{G}$ taken together must be equal to the number of edges in K_n.

Since K_n has $\dfrac{n(n-1)}{2}$ edges, it follows that each of G and $\overline{G}$ has $\dfrac{n(n-1)}{4}$ edges.

Thus, $\dfrac{n(n-1)}{4}$ must be a positive integer. That is n or $(n-1)$ must be a multiple of 4.

1.6.4   Union, Intersection, Ring sum and Product of Graphs

Definition 1.22 (Union). *The union of two graphs $G_1 = (V_1, E_1)$ and $G_2 = (V_2, E_2)$ is the simple graph with vertex set $V_1 \cup V_2$ and edge set $E_1 \cup E_2$. The union of G_1 and G_2 is denoted by $G_1 \cup G_2$.*

Definition 1.23 (Intersection). *The intersection of two graphs $G_1 = (V_1, E_1)$ and $G_2 = (V_2, E_2)$ is the simple graph with vertex set $V_1 \cap V_2$ and edge set $E_1 \cap E_2$. The union of G_1 and G_2 is denoted by $G_1 \cap G_2$.*

Example: We have

$V_1 = \{v_1, v_2, v_3, v_4, v_5\}$ $\qquad\qquad$ $V_2 = \{v_1, v_2, v_3, v_4, v_5, v_6, v_7\}$

$E_1 = \{e_1, e_2, e_3, e_4, e_5, e_6\}$ $\qquad\qquad$ $E_2 = \{e_2, e_3, e_5, e_6, e_7, e_8, e_9, e_{10}, e_{11}, e_{12}\}$

$V_1 \cup V_2 = \{v_1, v_2, v_3, v_4, v_5, v_6, v_7\}$ $\qquad$ $V_1 \cap V_2 = \{v_1, v_2, v_3, v_4, v_5\}$

$E_1 \cup E_2 = \{e_1, e_2, e_3, e_4, e_5, e_6, e_7, e_8, e_9, e_{10}, e_{11}, e_{12}\}$ $\qquad$ $E_1 \cap E_2 = \{e_2, e_3, e_5, e_6\}$

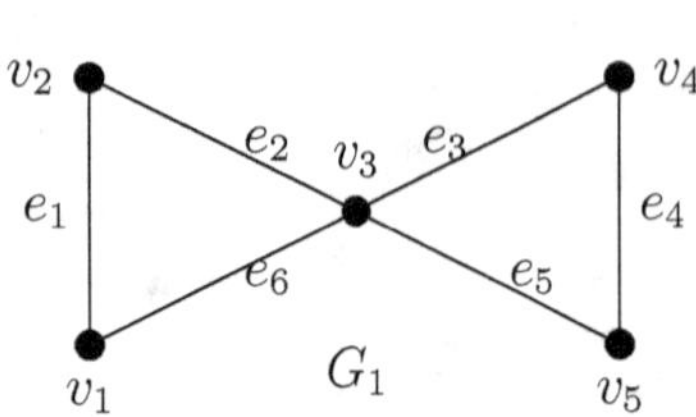
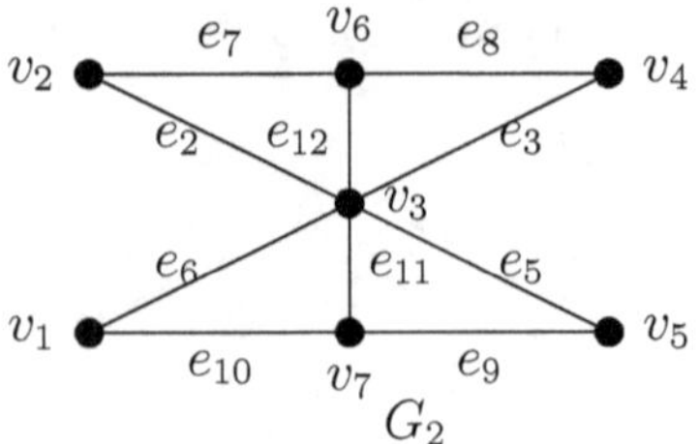

Figure 1.45

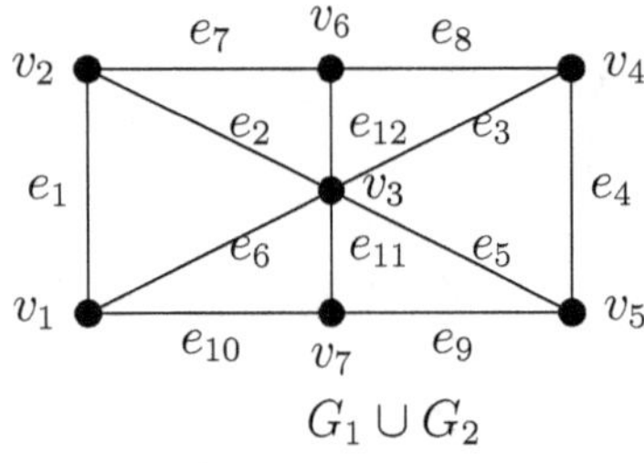
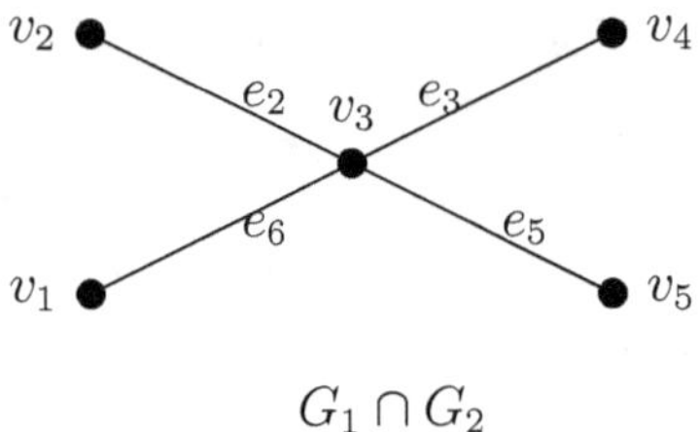

Figure 1.46

Definition 1.24 (Ring sum of two graphs). *The ring sum of two graphs $G_1 = (V_1, E_1)$ and $G_2 = (V_2, E_2)$ is a graph with vertex set $V = V_1 \cup V_2$ and edge set E consists of those edges which are in E_1 or E_2 but not in both. That is $E = E_1 \cup E_2 - (E_1 \cap E_2)$ The ring sum is denoted by $G_1 \oplus G_2$.*

For example:

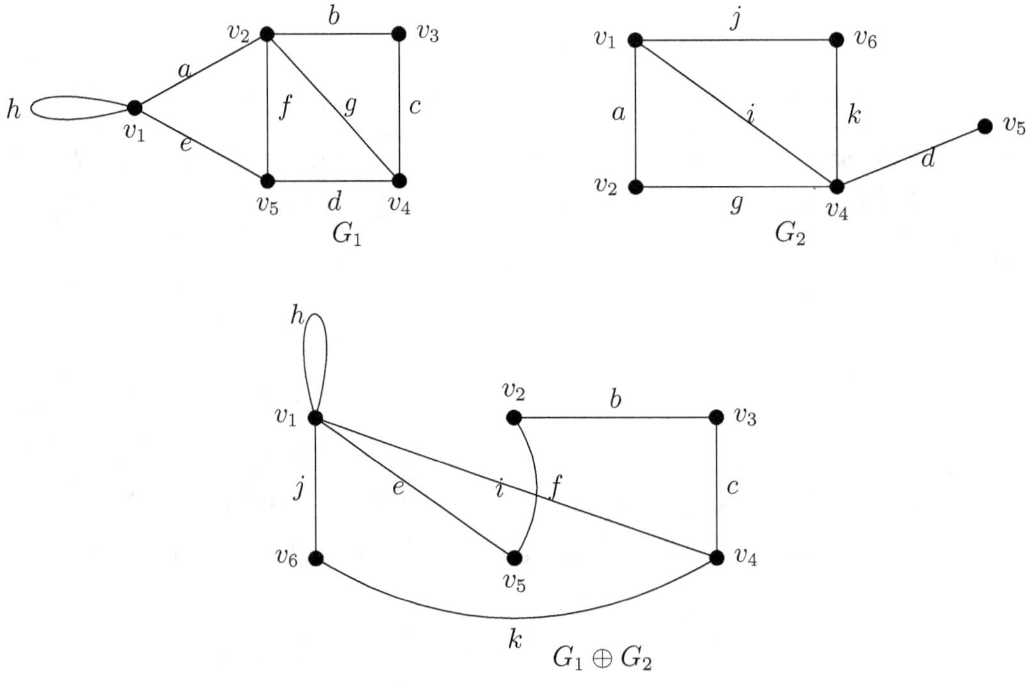

Definition 1.25 (Product of two graphs). *Let $G_1 = (V_1, E_1)$ and $G_2 = (V_2, E_2)$ be two simple graphs with $V_1 \cap V_2 = \phi$. The product of G_1 and G_2 denoted by $G_1 \times G_2$ is the graph having vertex set as $V = V_1 \times V_2$. Then $u = (u_1, u_2)$ and $v = (v_1, v_2)$ are adjacent in $G_1 \times G_2$ whenever $u_1 = v_1$ and u_2 adjacent to v_2 or $u_2 = v_2$ and u_1 adjacent to v_1.*

For example, consider following graph

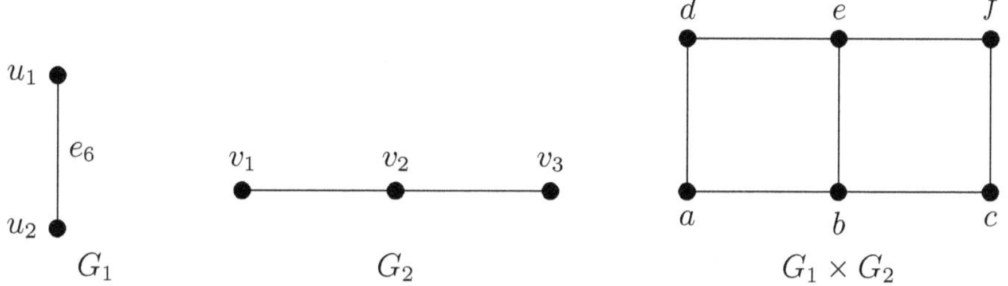

Figure 1.48

where $a = (u_1, v_1)$, $b = (u_1, v_2)$, $c = (u_1, v_3)$, $d = (u_2, v_1)$, $e = (u_2, v_2)$, $f = (u_2, v_3)$.

Definition 1.26 (Fusion of vertices).

Let $G(V, E)$ be a graph and v_1 and v_2 are two vertices in G then fusion of these two vertices means they are repalced by a single vertex (at v_1 or v_2) in such a manner that all the edges in G which were incident on v_1 and v_2 are now made incident on that single vertex.

Thus fusion of two vertices reduces the number of vertices by 1 but it does not alter the number of edges.

For example: In the following graph G fuse the vertices u_1 and u_2.

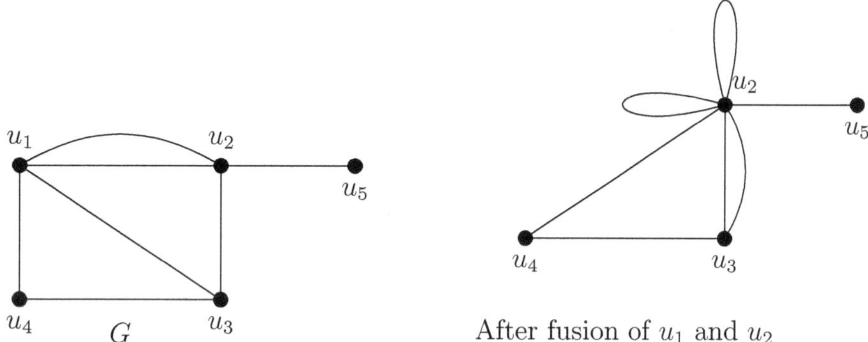

After fusion of u_1 and u_2

★ <u>**Illustrative Examples**</u> ★

Example 1.27. *Find the complement of the following graphs.*

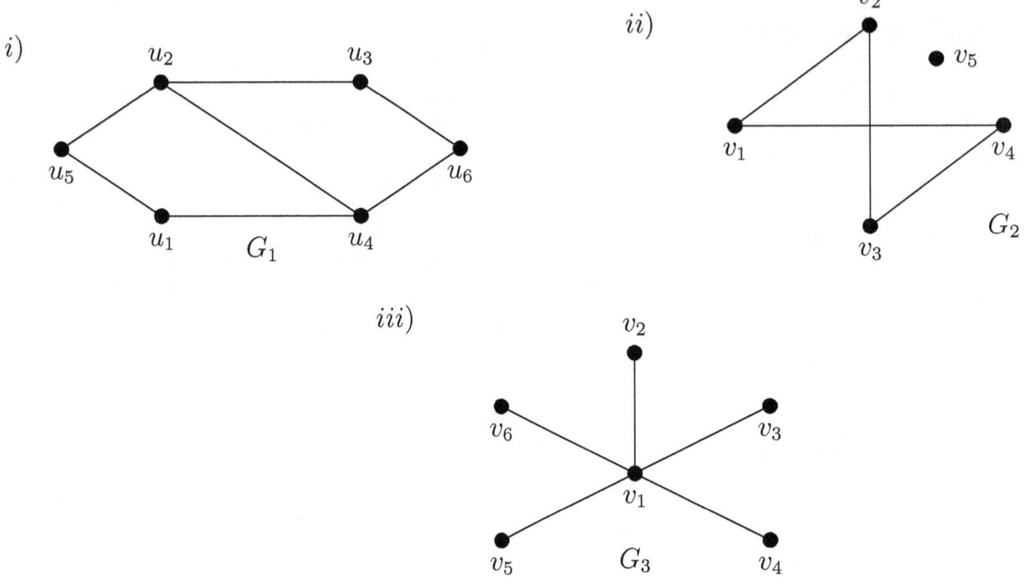

Figure 1.49

Solution:

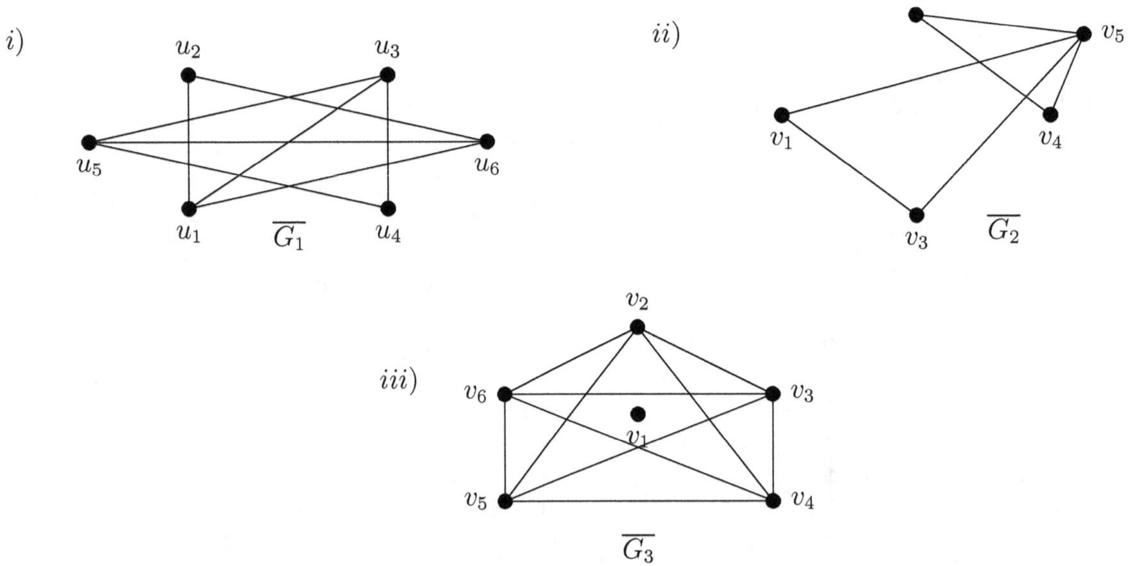

Figure 1.50

Example 1.28. *Consider the following graph G.*

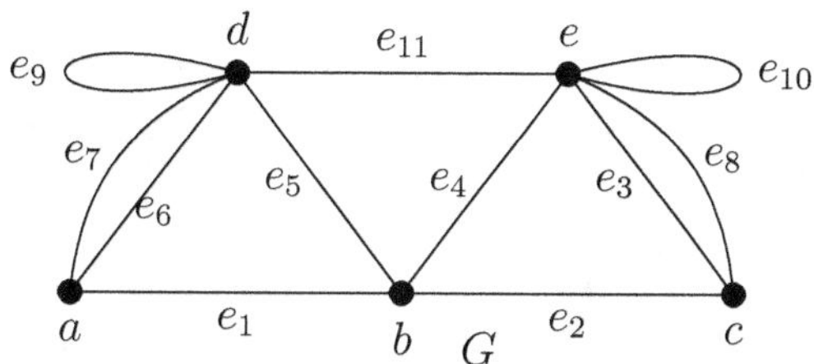

a) *Find induced subgraph $G < V >$ where $V = \{a, d, e, c\}$*

b) *Find $G - A$ where $A = \{e_3, e_6, e_9, e_{10}\}$.*

c) *The new graph G_1 obtained from G by fusion of vertices d and e .*

Solution:

a) Induced subgraph $G < V >$ where $V = \{a, d, e, c\}$

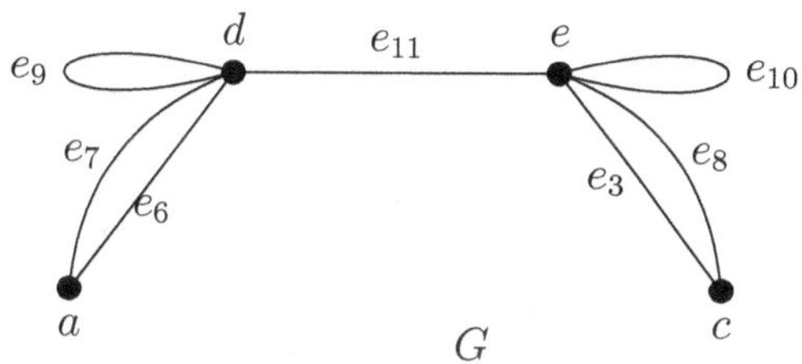

b) $G - A$ where $A = \{e_3, e_6, e_9, e_{10}\}$.

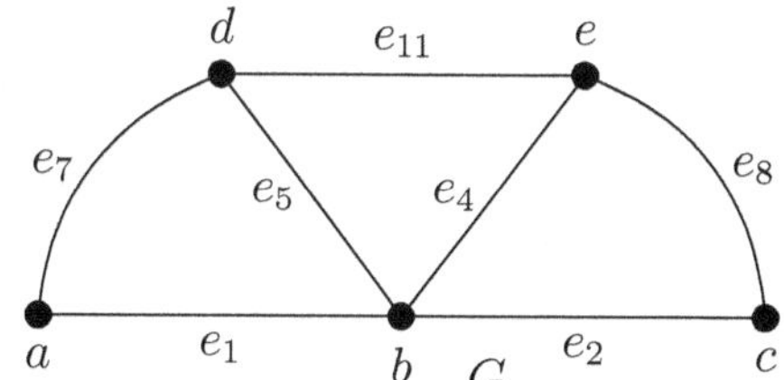

c) Fusion of vertices d and e

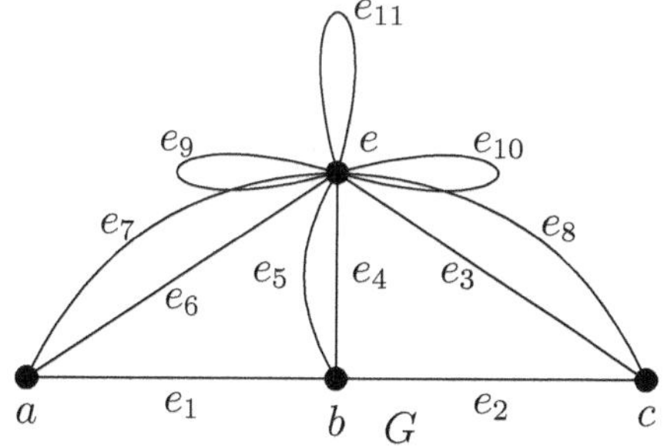

Example 1.29. *For the given graphs G_1 and G_2, find $G_1 \cup G_2$ and $G_1 \cap G_2$.*

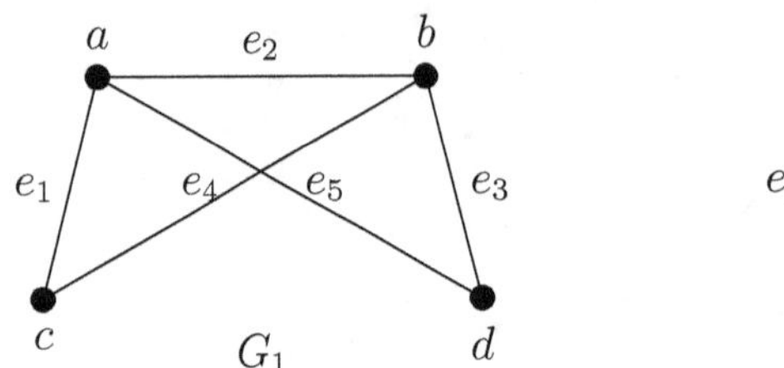
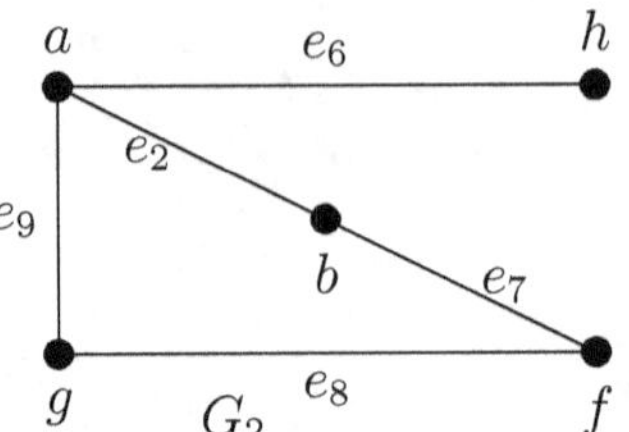

Figure 1.51

Solution: We have

$$V_1 = \{a, b, c, d\} \qquad\qquad V_2 = \{a, b, h, g, f\}$$

$$E_1 = \{e_1, e_2, e_3, e_4, e_5\} \qquad\qquad E_2 = \{e_2, e_6, e_7, e_8, e_9\}$$

$$V_1 \cup V_2 = \{a, b, c, d, h, g, f\} \qquad\qquad V_1 \cap V_2 = \{a, b\}$$

$$E_1 \cup E_2 = \{e_1, e_2, e_3, e_4, e_5, e_6, e_7, e_8, e_9\} \qquad E_1 \cap E_2 = \{e_2\}$$

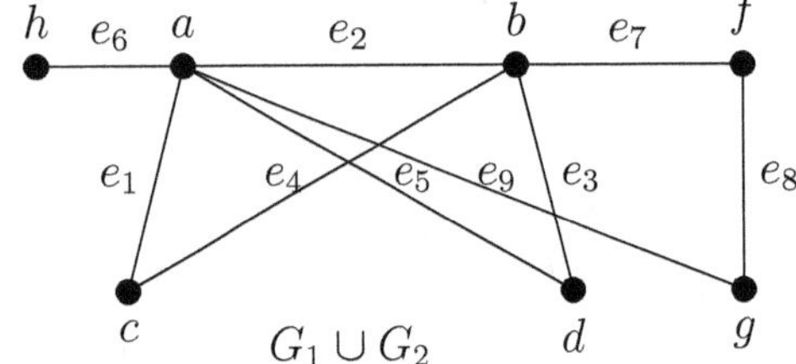
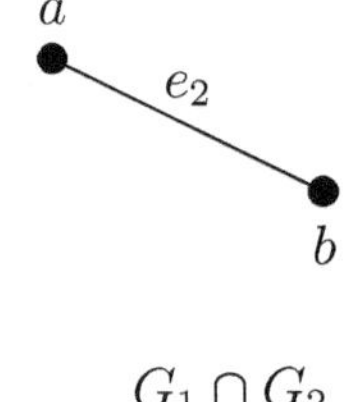

Example 1.30. *For the given graphs G_1 and G_2, find $G_1 \cup G_2$, $G_1 \cap G_2$ and $G_1 \oplus G_2$.*

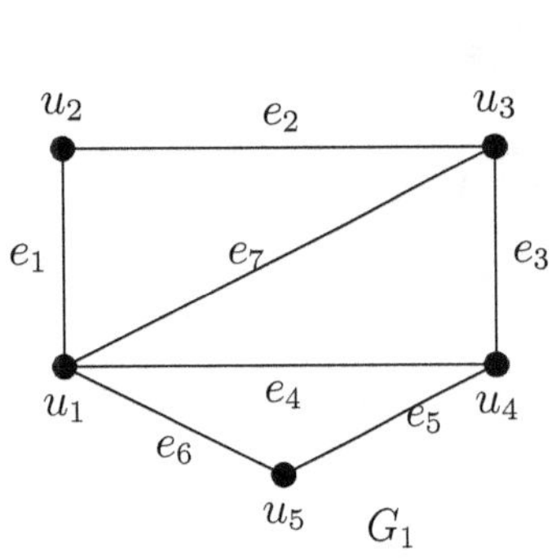
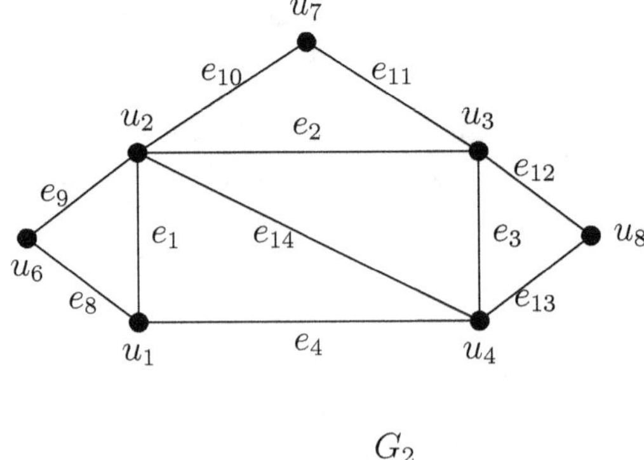

Figure 1.52

Solution: We have

$$V_1 = \{u_1, u_2, u_3, u_4, u_5\}, \qquad V_2 = \{u_1, u_2, u_3, u_4, u_6, u_7, u_8\}$$

$$E_1 = \{e_1, e_2, e_3, e_4, e_5, e_6, e_7\}, \qquad E_2 = \{e_1, e_2, e_3, e_4, e_8, e_9, e_{10}, e_{11}, e_{12}, e_{13}, e_{14}\}$$

$$V_1 \cup V_2 = \{u_1, u_2, u_3, u_4, u_5, u_6, u_7, u_8\}, \qquad V_1 \cap V_2 = \{u_1, u_2, u_3, u_4\}$$

$$E_1 \cup E_2 = \{e_1, e_2, e_3, e_4, e_5, e_6, e_7, e_8, e_9, e_{10}, e_{11}, e_{12}, e_{13}, e_{14}\}, \qquad E_1 \cap E_2 = \{e_1, e_2, e_3, e_4\}$$

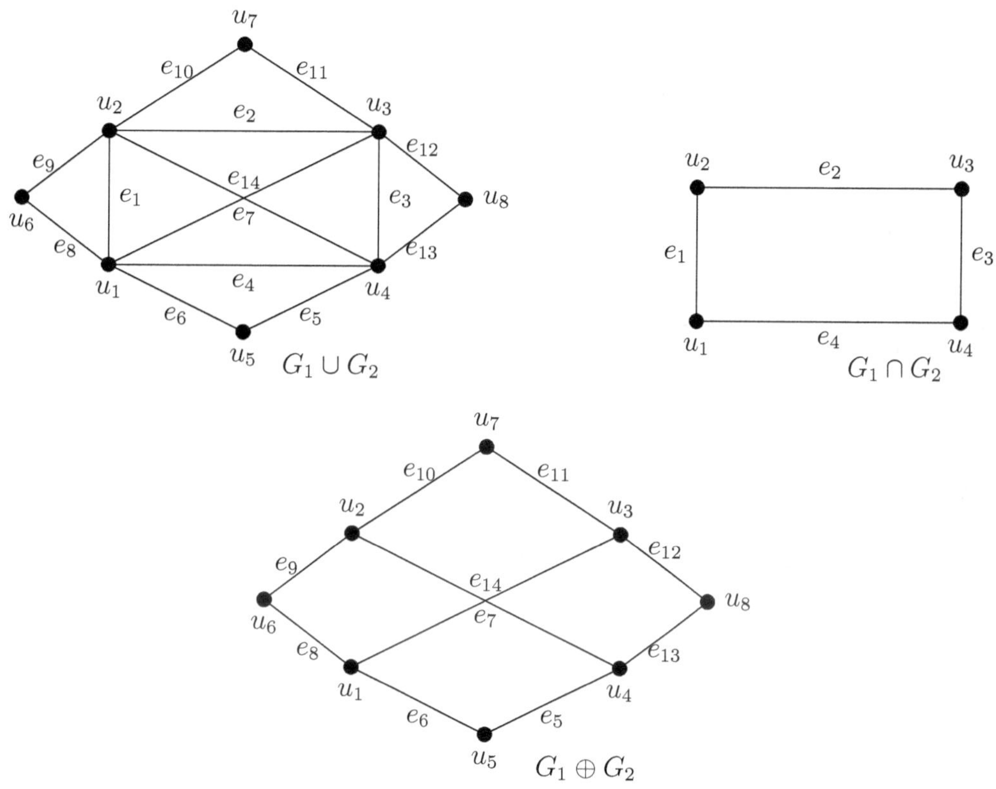

Example 1.31. *For the given graphs G_1 and G_2, find $G_1 \times G_2$.*

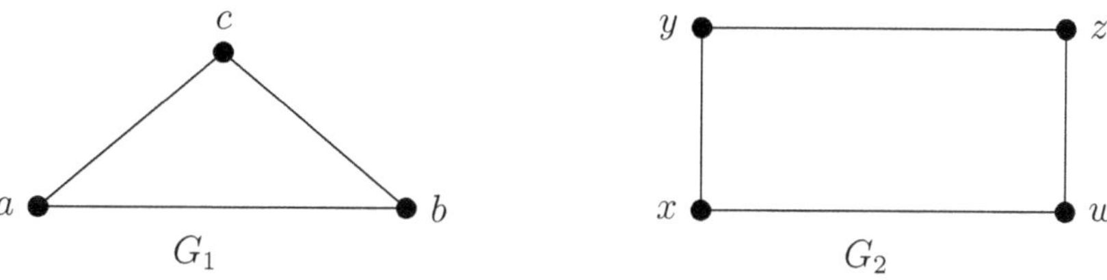

Solution: Let $u_1 = (a, x),\ u_2 = (a, y),\ u_3 = (a, z),\ u_4 = (a, w),$

$$u_5 = (b, x),\ u_6 = (b, y),\ u_7 = (b, z),\ u_8 = (b, w),$$

$$u_9 = (c, x),\ u_{10} = (c, y),\ u_{11} = (c, z),\ u_{12} = (c, w).$$

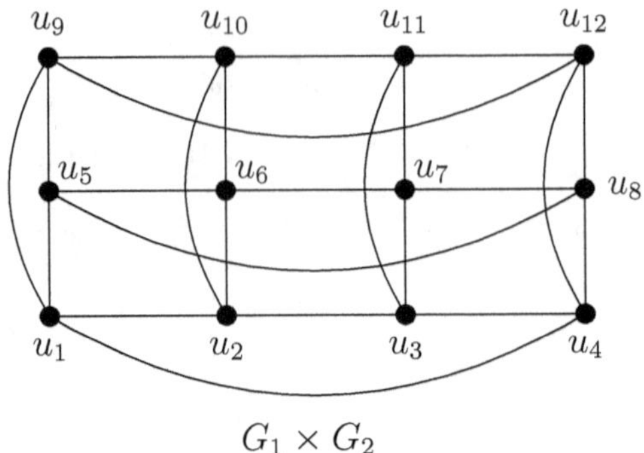

$$G_1 \times G_2$$

Example 1.32. *For the given graphs G_1 and G_2, find $G_1 \cup G_2$ and $G_1 \oplus G_2$.*

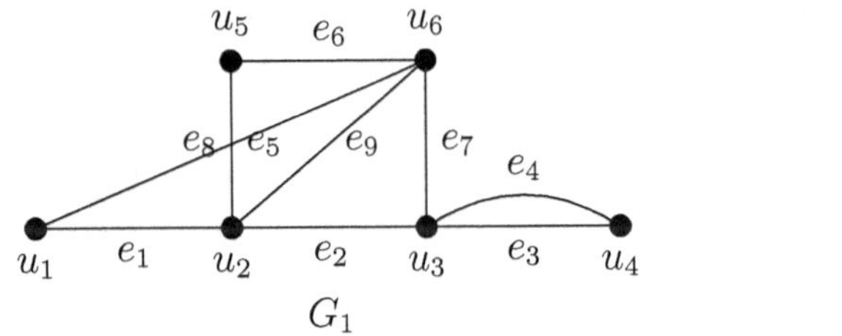

$$G_1$$

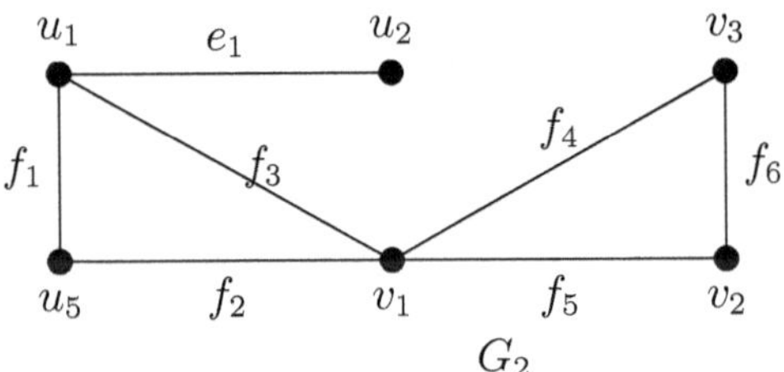

$$G_2$$

Solution: We have

$$V_1 = \{u_1, u_2, u_3, u_4, u_5, u_6\} \qquad\qquad V_2 = \{u_1, u_2, u_5, v_1, v_2, v_3\}$$

$$E_1 = \{e_1, e_2, e_3, e_4, e_5, e_6, e_7, e_8, e_9\} \qquad\qquad E_2 = \{e_1, f_1, f_2, f_3, f_4, f_5, f_6\}$$

$$V_1 \cup V_2 = \{u_1, u_2, u_3, u_4, u_5, u_6, v_1, v_2, v_3\} \qquad V_1 \cap V_2 = \{u_1, u_2, u_5\}$$

$$E_1 \cup E_2 = \{e_1, e_2, e_3, e_4, e_5, e_6, e_7, e_8, e_9, f_1, f_2, f_3, f_4, f_5, f_6\} \qquad E_1 \cap E_2 = \{e_1\}$$

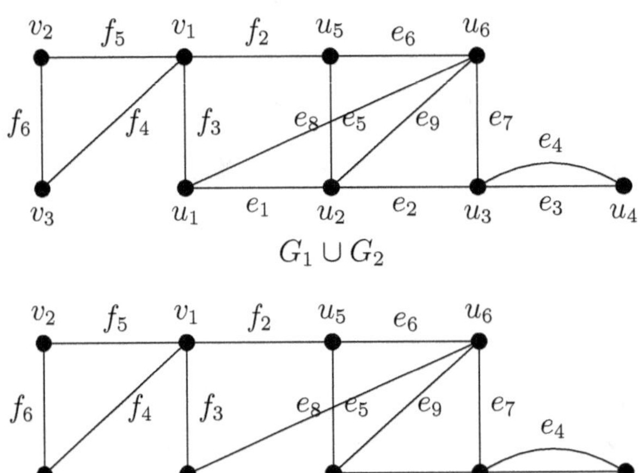

$$G_1 \cup G_2$$

$$G_1 \oplus G_2$$

Figure 1.53

Exercise: 1.3

1. List all self complementary graphs on 4 vertices.

2. Draw the complement of following graphs.

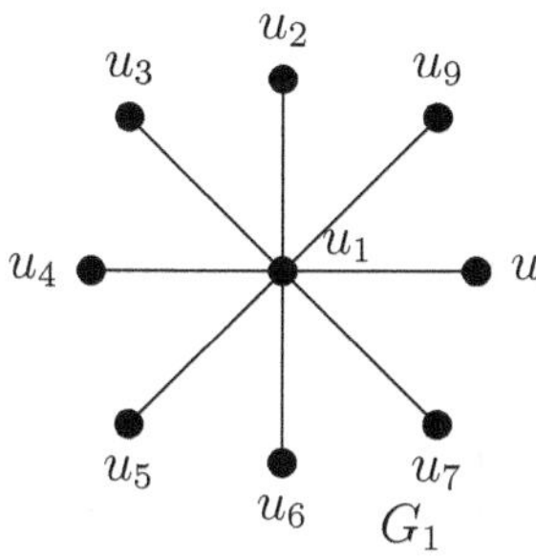 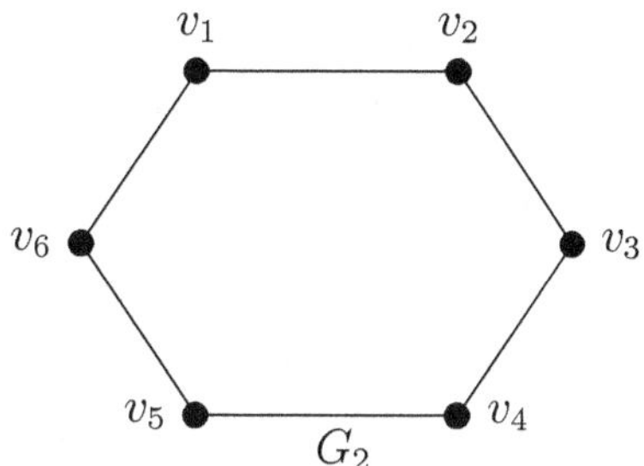

3. Find all nonisomorphic subgraphs of K_4.

4. For the given grpah G, Find

    i) underlying graph of G   ii) $G - A; A = \{v_3, v_5\}$

    iii) $G - B; B = \{e_1, e_3, e_9, e_{10}, e_{11}, e_{12}\}$ iv)$< H >; H = \{v_3, v_4, v_5, v_6\}$

    v) $< F >; F = \{e_{13}, e_{14}, e_{15}, e_{16}\}$

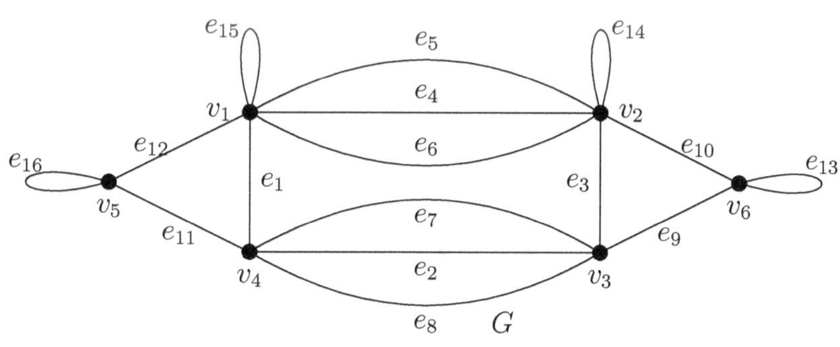

5. Find $G_1 \cup G_2$, $G_1 \cap G_2$ and $G_1 \oplus G_2$ for the following pair of graphs.

    i)

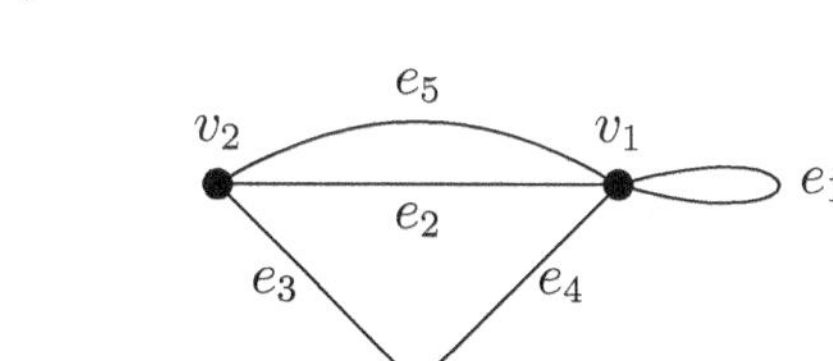 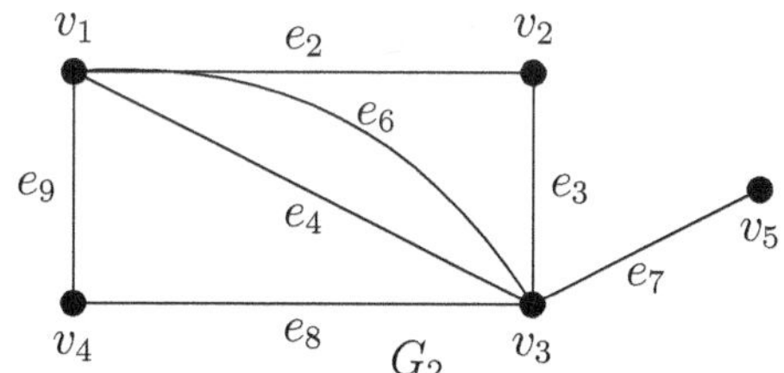

ii)

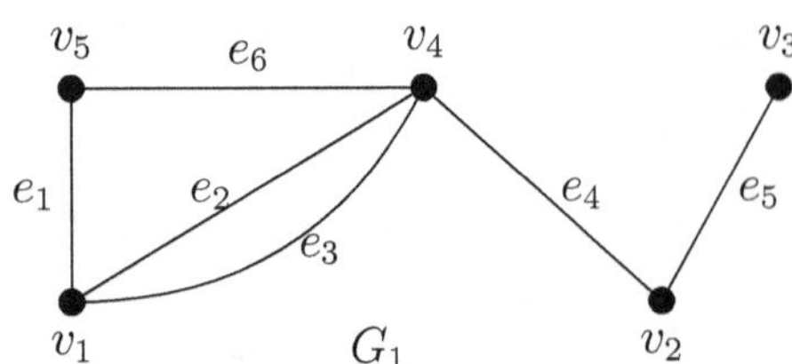

G_1 G_2

6. For the given graphs G_1 and G_2, find $G_1 \times G_2$.

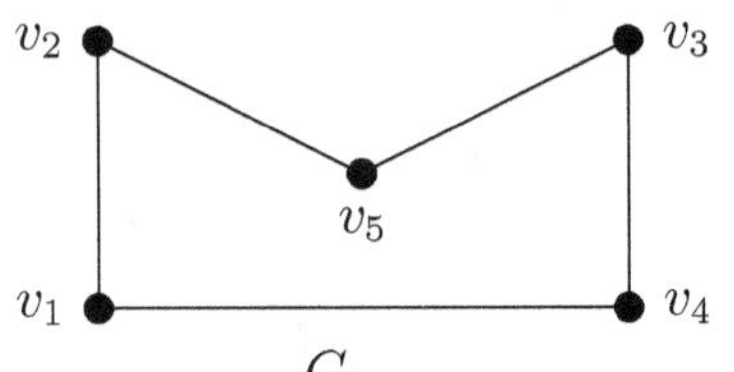

G_1

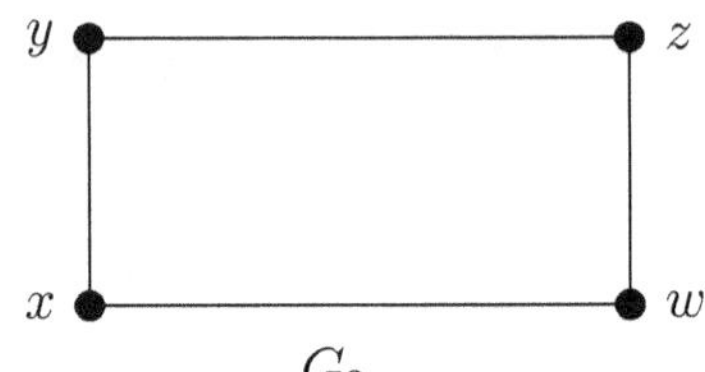

G_2

Solutions

1.

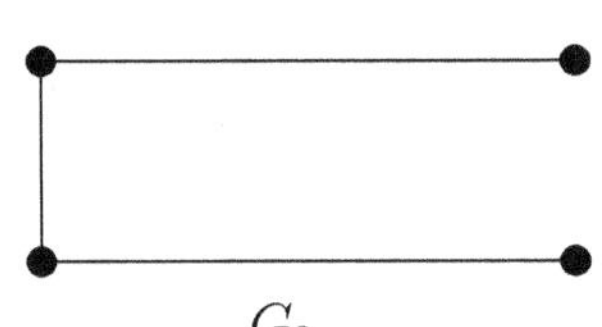

G_2

2.

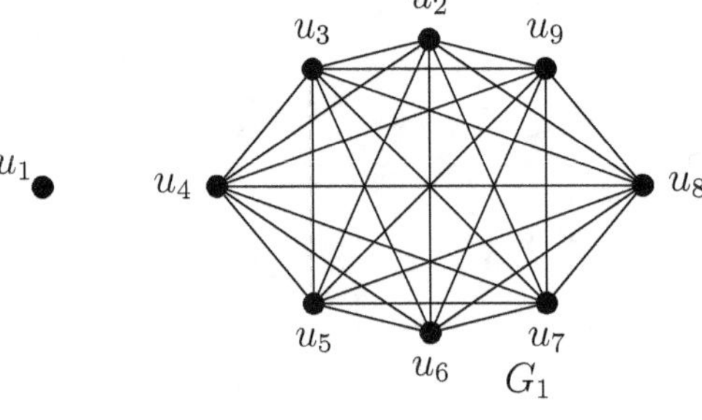

G_1

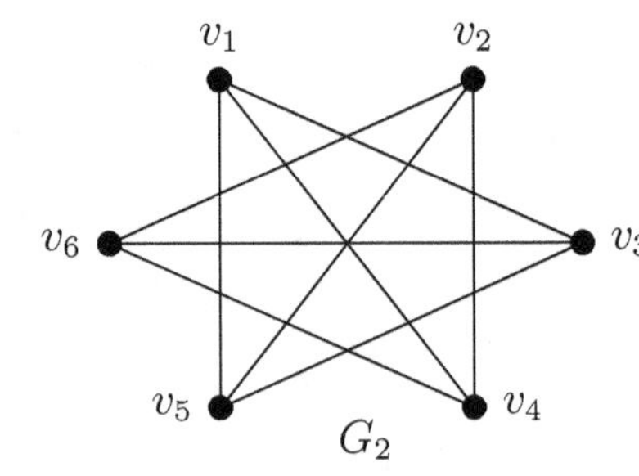

G_2

3.

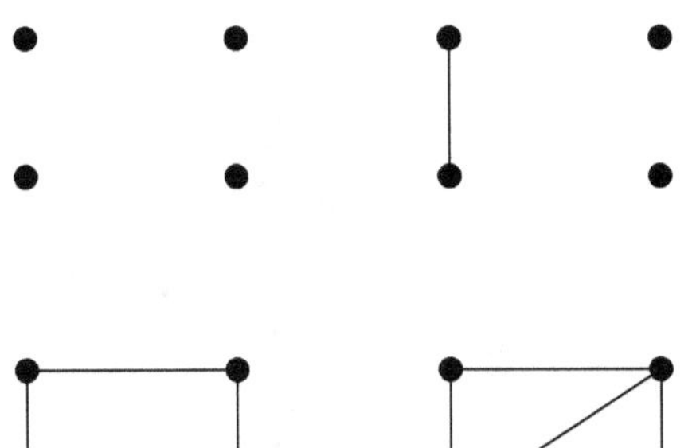

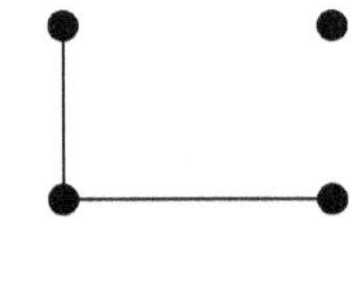

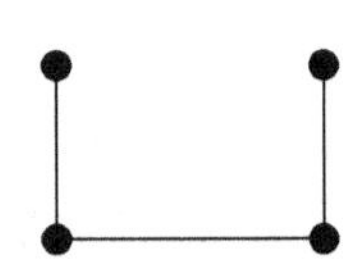

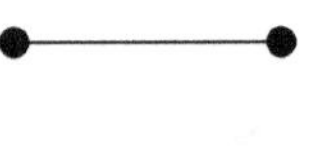

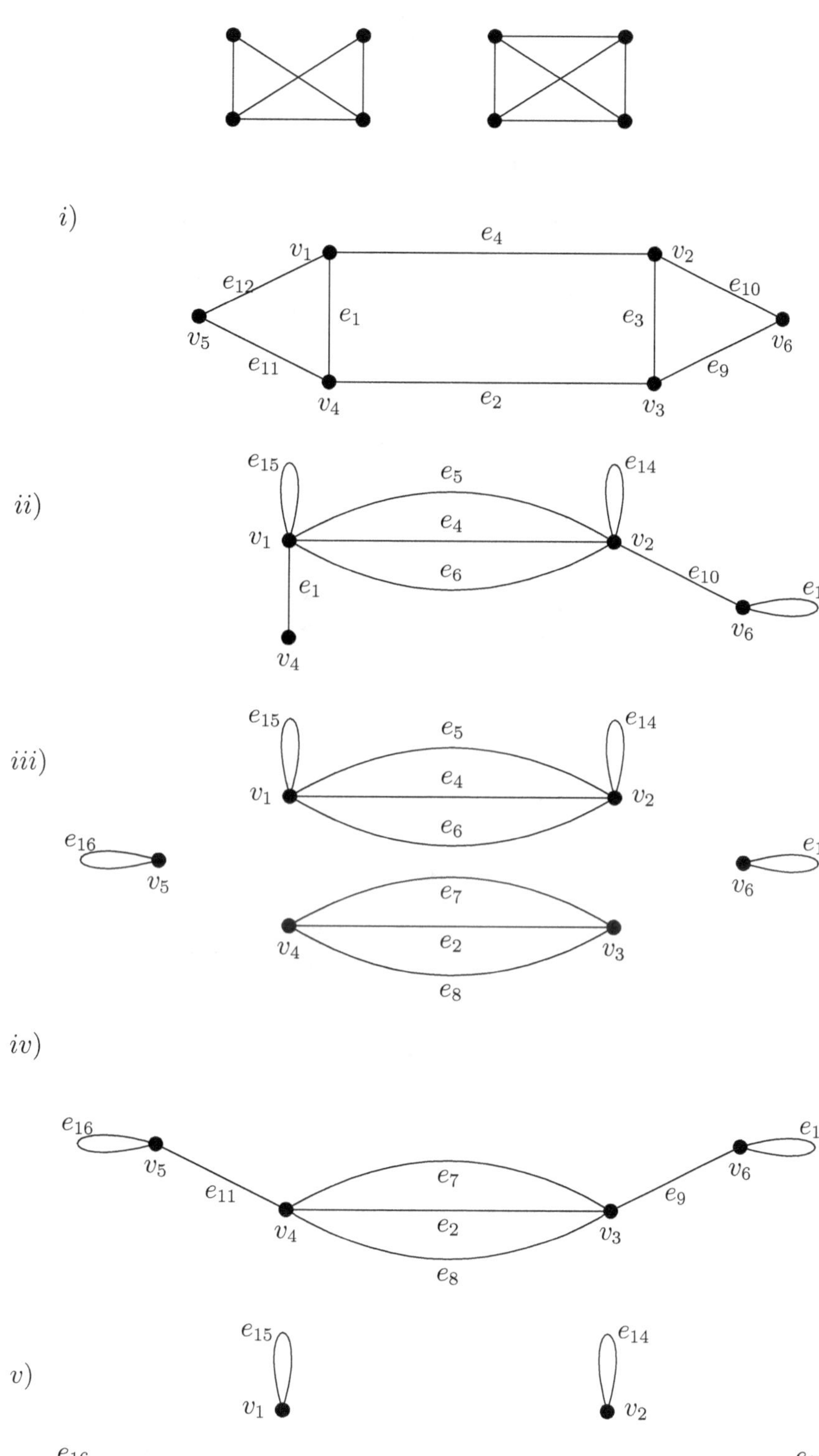

4.

i)

ii)

iii)

iv)

v)

5. i)

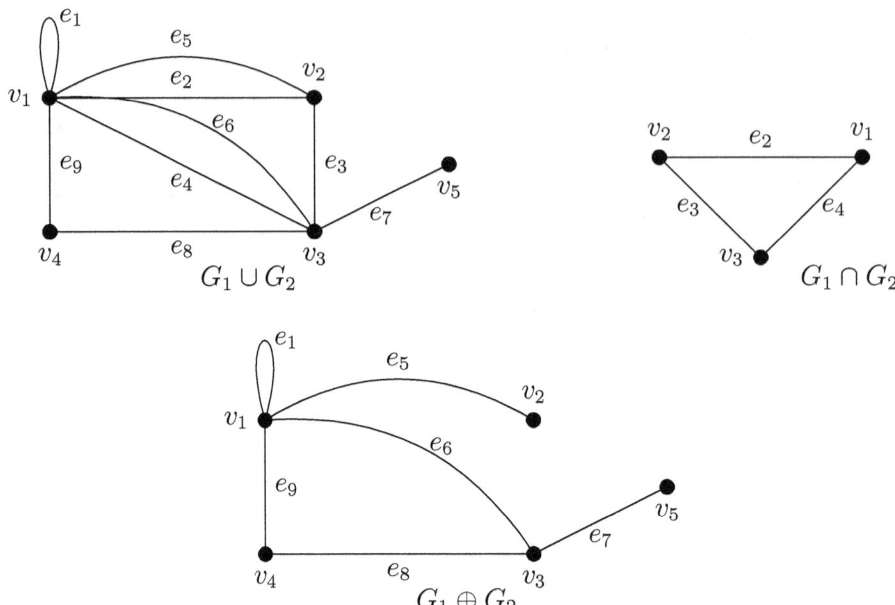

6.

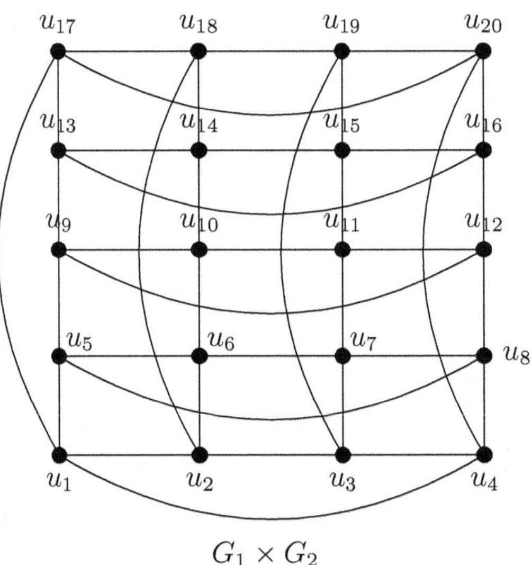

where

$$u_1 = (v_1, x), \ u_2 = (v_1, y), \ u_3 = (v_1, z), \ u_4 = (v_1, w),$$
$$u_5 = (v_2, x), \ u_6 = (v_2, y), \ u_7 = (v_2, z), \ u_8 = (v_2, w),$$
$$u_9 = (v_3, x), \ u_{10} = (v_3, y), \ u_{11} = (v_3, z), \ u_{12} = (v_3, w).$$
$$u_{13} = (v_4, x), \ u_{14} = (v_4, y), \ u_{15} = (v_4, z), \ u_{16} = (v_4, w).$$
$$u_{17} = (v_5, x), \ u_{18} = (v_5, y), \ u_{19} = (v_5, z), \ u_{20} = (v_5, w).$$

Chapter 2

Connected graphs

In Chapter 1, we defined basic terminology of graphs and introduced concepts like matrix representations of graphs, operations on graphs and isomorphism. In this chapter we will study concepts like walk, path, cycle etc. which would help us to understand the definition of connected graph. We also study isthmus and cut vertex and algorithm to find shortest path between two vertices.

2.1 Walk, Path and Circuits

Definition 2.1. *Let G be a graph. A **walk** in a graph G is defined as a finite alternating sequences of vertices and edges $v_0, e_1, v_1, e_2, v_2, e_3, \cdots, v_{n-1}, e_n, v_n$ beginning and ending with vertices in which each edge is incident with two vertices immediately proceeding and following it.*

A walk in a graph is denoted by w. The walk $v_0, e_1, v_1, e_2, v_2, e_3, \cdots, v_{n-1}, e_n, v_n$ may be written as $v_0 - v_1 - v_2, - \cdots, v_{n-1} - v_n$ and is called $v_0 - v_n$ walk. The vertices v_0 and v_n with which the walk begins and ends are called the initial and the terminal vertices. If $v_0 \neq v_n$ then the walk is called an **open walk.** $v_0 = v_n$ then the walk is called a **closed walk.**

Definition 2.2. *A walk in a graph; in which no edge is repeated is called a **trail.***

Definition 2.3. *A closed trail is called **tour** in a graph G.*

Definition 2.4. *The number of edges in walk is called the length of the walk.*

Example 2.1. *Consider the graph given below*

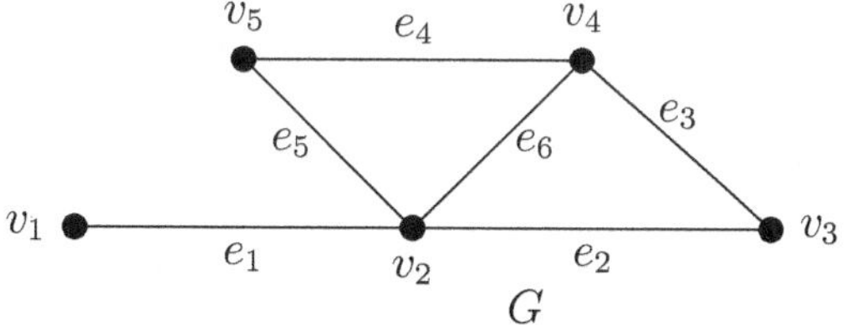

Figure 2.1

$W_1 : v_1\ e_1\ v_2\ e_2\ v_3\ e_3\ v_4\ e_4\ v_5\ e_4\ v_4\ e_6\ v_2$ is a walk.

$W_2 : v_1\ e_1\ v_2\ e_2\ v_3\ e_3\ v_4\ e_4\ v_5.$

Since the edges are not repeated the walk W_2, is a trail.

$W_3 : v_1\ e_1\ v_2\ e_2\ v_3\ e_3\ v_4\ e_6\ v_2\ e_1\ v_1$ is a closed walk.

$W_4 : v_2\ e_2\ v_3\ e_3\ v_4\ e_6\ v_2$ is closed trail, hence a tour.

Definition 2.5. *A walk in which any vertex is non-repeated (hence there cannot be repetition of any edge) is called a **path.***

In Figure 2.1, $C : v_1\ e_1\ v_2\ e_2\ v_3\ e_3\ v_4$ is a path.

Definition 2.6. *A closed walk in which no vertex (except its terminal vertices) appear more than once is called a **circuit or cycle.***

In Figure 2.1, $P : v_2\ e_2\ v_3\ e_3\ v_4\ e_6\ v_2$ is a cycle.

Remark 2.1.

1) *The number of edges in a path (cycle) is called length of that path (cycle).*

2) *A path of length greater than zero that begins and ends at the same vertex is known as circuit or cycle*

3) *Self loop is a circuit.*

4) *A regular graph of degree 2 is a circuit.*

Theorem 2.1. *In a graph G there exists a path from the vertex u to the vertex v if and only if there exists a walk from u to v.*

Proof: We know that every path is a walk. Therefore, if there is a path from u to v then it can be considered as a walk from u to v.

Conversly suppose that there is a walk W from u to v, i.e. $W : u\ e_1\ v_1\ e_2\ \cdots\ v$. If in W each vertex appears only once then W is itself a path from u to v. If not then some vertices in W are repeated. Let v_i be the first vertex which is repeated in W. Therefore walk W is of the form

$$W : u\ e_1\ v_1\ \cdots\ e_{i-1}\ v_{i-1}\ e_i\ v_i\ e_{i+1}\ v_{i+1}\ \cdots\ e_j\ v_i\ \cdots\ v$$

From W we delete the portion $e_{i+1}\ v_{i+1}\ \cdots\ e_j$. Then W reduces to

$$W' : u\ e_1\ v_1\ \cdots\ e_{i-1}\ v_{i-1}\ e_i\ v_i\ \cdots\ v$$

If there is no repetition of any vertex in W' then W' is a path from u to v. Otherwise suppose that the vertex v_j is the first vertex in W' which is repeated. As before we delete the portion $e_{j+1}\ v_{j+1}\ \cdots\ e_k$ from W' and get W''. We continue this process of deletion until we get W^* where

2.2　Connected Graph

Definition 2.7. *An undirected graph G is called connected if there is a path between every pair of distinct vertices of the graph.*

If G is not connected then it is said to be disconnected. In a disconnected graph, there is atleast one pair of vertices which is not connected.

Consider the graphs given below.

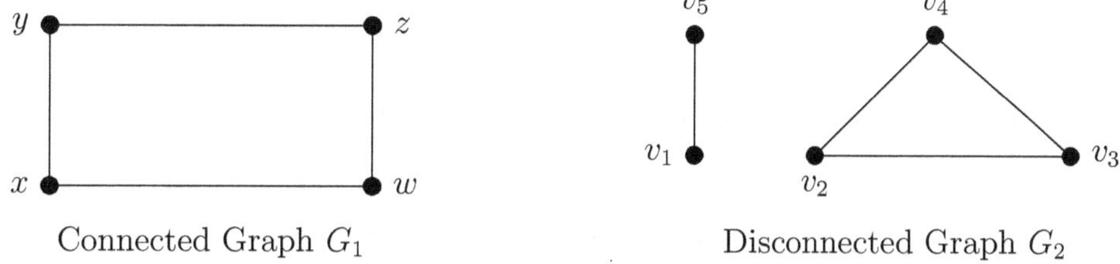

Connected Graph G_1　　　　　　　　　　　Disconnected Graph G_2

Figure 2.2

A graph G_1 is connected with 4 vertices and 4 edges. In graph G_2, there is no path from v_1 to v_2. Therefore G_2 is disconnected graph with 5 vertices and 4 edges.

Definition 2.8. *A connected subgraph H of a disconnected graph G, which is not properly contained in any proper connected subgraph of G is called maximal connected subgraph or component of G.*

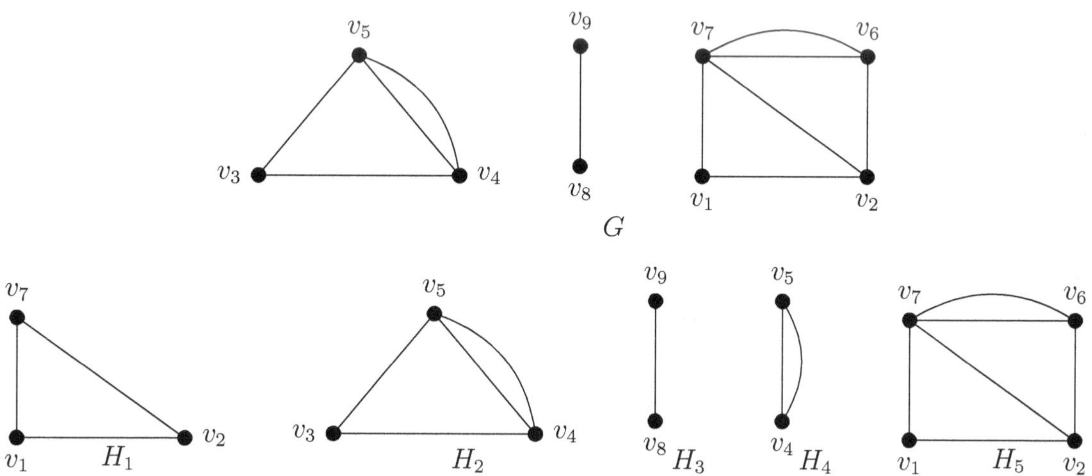

Figure 2.3

H_2, H_3, and H_5 are components of G but H_1 and H_4 are not because H_1 is a subgraph of H_5 and H_4 is a subgraph of H_2.

Theorem 2.2. *A graph G is disconnected if and only if its vertex set V can be partitioned into two subsets V_1 and V_2 such that there exists no edge in G whose one end vertex is in the subset V_1 and*

Proof: Let G be disconnected. Then we have by the definition that there exists a vertex x in G and a vertex y in G such that there is no path between x and y in G Let $V_1 = \{z \in V : z \text{ is connected to x}\}$. Then V_1 is the set of all vertices of G which are connected to x. Let $V_2 = V - V_1$. Then $V_1 \cap V_2 = \phi$ and $V_1 \cup V_2 = V$.

Hence V_1 and V_2 are the partition of $V(G)$. Let a be any vertex of V_1. To prove that a is not adjacent to any vertex of V_2. If possible let $b \in V_2$ such that $ab \in E(G)$. Then $a \in V_1$ there exist a path P_1 : from x to a. This path can be extended to the path $P_2 = P_1, ab$. Then P_2 is a path from x to b in G. Therefore x and b are connected. This implies that $b \in V_1$ which is contradiction to the fact $V_1 \cap V_2 = \phi$.

Conversely, let us assume that V can be partitioned into two subsets V_1 and V_2 such that no vertex of V_1 is adjacent to a vertex of V_2.

Let x be any vertex in V_1 and y be any vertex in V_2. Then there is no path between x and y. To prove that G is disconnected, if possible, suppose G is connected. Then x and y are connected. Therefore, there exists a path between x and y in G. But this path is possible only through a vertex W in G which is not either in V_1 or V_2. Hence $V_1 \cup V_2 \neq V$, a contradiction.

Theorem 2.3. *A simple graph with n vertices and k components cannot have more than $\dfrac{(n-k)(n-k+1)}{2}$ edges.*

Proof: Let G Let G be a simple graph with n vertices and k components. If $G_1, G_2, \cdots, G_k$ are k components of G having number of vertices $n_1, n_2, \cdots, n_k$ respectively then we have

$$n_1 + n_2 + \cdots + n_k = n. \text{ i.e. } \sum_{j=1}^{k} n_j = n. \text{ Therefore } \sum_{j=1}^{k}(n_j - 1) = \sum_{j=1}^{k} n_j - \sum_{j=1}^{k} 1 = n - k.$$

Consider

$$\left[\sum_{j=1}^{k}(n_j - 1)\right]^2 = (n - k)^2$$

$$\sum_{j=1}^{k}(n_j - 1)^2 + \sum_{i \neq j}(n_i - 1)(n_j - 1) = (n - k)^2$$

$$\sum_{j=1}^{k}(n_j - 1)^2 + \text{sum of non-negative terms} = n^2 - 2nk + k^2$$

$$\sum_{j=1}^{k}(n_j^2 - 2n_j + 1) \leq n^2 - 2nk + k^2$$

$$\sum_{j=1}^{k} n_j^2 - 2\sum_{j=1}^{k} n_j + \sum_{j=1}^{k} 1 \leq n^2 - 2nk + k^2$$

$$\Rightarrow \qquad \sum_{j=1}^{k} n_j^2 \leq n^2 - 2nk + k^2 + 2n - k. \qquad (2.1)$$

in G_j is $\frac{1}{2}n_j(n_j - 1)$. Hence, the total number of edges in G is

$$
\begin{aligned}
&\leq \frac{1}{2}\sum_{j=1}^{k} n_j(n_j - 1) = \frac{1}{2}\left(\sum_{j=1}^{k} n_j^2 - \sum_{j=1}^{k} n_j\right) = \frac{1}{2}\left(\sum_{j=1}^{k} n_j^2 - n\right) \\
&\leq \frac{1}{2}\left(n^2 - 2nk + k^2 + 2n - k - n\right). \qquad \text{by (2.1)} \\
&= \frac{1}{2}\left(n^2 - 2nk + k^2 + n - k\right) \\
&= \frac{1}{2}\left((n - k)^2 + (n - k)\right) \\
&= \frac{(n - k)(n - k + 1)}{2}. \qquad \text{Hence proved.}
\end{aligned}
$$

2.3 Counting paths between vertices

The number of paths between two vertices in a graph can be determined using its adjacency matrix.

Theorem 2.4. *Let G be a graph with adjacency matrix A with respect to the ordering v_1, v_2, $\cdots$, v_n (with directed or undirected edges, with multiple edges and loops allowed). The number of different paths of length r ($r \in \mathbb{N}$) from v_i to v_j is equal to the $(i,j)^{th}$ entry of A^r.*

Example 2.2. *In the following graph find*
a) number of paths of length 2 form v_1 to v_4
b) number of paths of length 3 form v_2 to v_5.

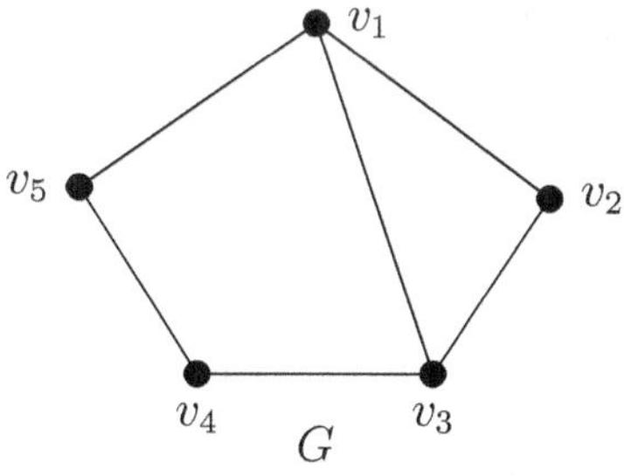

Figure 2.4

Solution : The adjacency matrix of G is

$$
A = \begin{array}{c}
\\ v_1 \\ v_2 \\ v_3 \\ v_4 \\ v_5
\end{array}
\begin{array}{c}
\begin{array}{ccccc} v_1 & v_2 & v_3 & v_4 & v_5 \end{array} \\
\left[\begin{array}{ccccc}
0 & 1 & 1 & 0 & 1 \\
1 & 0 & 1 & 0 & 0 \\
1 & 1 & 0 & 1 & 0 \\
0 & 0 & 1 & 0 & 1 \\
1 & 0 & 0 & 1 & 0
\end{array}\right]
\end{array}
$$

a) for the paths of length 2, we find A^2.

$$A^2 = \begin{array}{c} \\ v_1 \\ v_2 \\ v_3 \\ v_4 \\ v_5 \end{array} \begin{array}{ccccc} v_1 & v_2 & v_3 & v_4 & v_5 \\ \begin{bmatrix} 3 & 1 & 1 & 2 & 0 \\ 1 & 2 & 1 & 1 & 1 \\ 1 & 1 & 3 & 0 & 2 \\ 2 & 1 & 0 & 2 & 0 \\ 0 & 1 & 2 & 0 & 2 \end{bmatrix} \end{array}$$

Number of paths of length 2 form v_1 to $v_4 = (1,4)^{th}$ entry $= 2$.

b) for the paths of length 3, we find A^3.

$$A^3 = \begin{array}{c} \\ v_1 \\ v_2 \\ v_3 \\ v_4 \\ v_5 \end{array} \begin{array}{ccccc} v_1 & v_2 & v_3 & v_4 & v_5 \\ \begin{bmatrix} 2 & 4 & 6 & 1 & 5 \\ 4 & 2 & 4 & 2 & 2 \\ 6 & 4 & 2 & 5 & 1 \\ 1 & 2 & 5 & 0 & 4 \\ 5 & 2 & 1 & 4 & 0 \end{bmatrix} \end{array}$$

Number of paths of length 3 form v_2 to $v_5 = (2,5)^{th}$ entry $= 2$.

Example 2.3. *find number of paths of length 4 form a to c in the directed graph given below.*

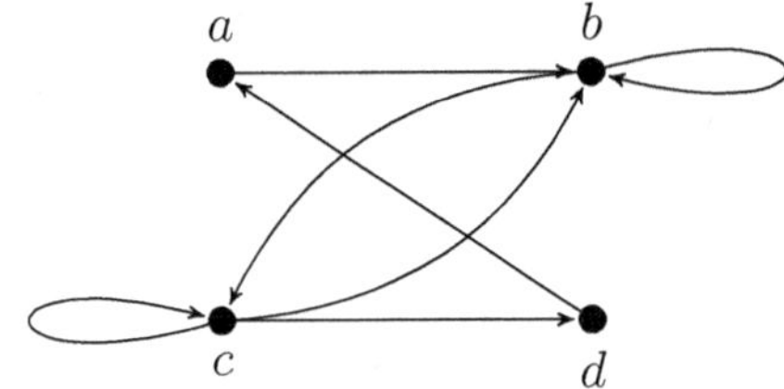

Solution : Adjacency matrix is

$$A = \begin{array}{c} \\ a \\ b \\ c \\ d \end{array} \begin{array}{cccc} a & b & c & d \\ \begin{bmatrix} 0 & 1 & 0 & 0 \\ 0 & 1 & 1 & 0 \\ 0 & 1 & 1 & 1 \\ 1 & 0 & 0 & 0 \end{bmatrix} \end{array}$$

Then

$$A^4 = \begin{array}{c} \\ a \\ b \\ c \\ d \end{array} \begin{array}{cccc} a & b & c & d \\ \begin{bmatrix} 1 & 4 & 4 & 2 \\ 2 & 9 & 8 & 4 \\ 2 & 10 & 9 & 4 \\ 0 & 2 & 2 & 1 \end{bmatrix} \end{array}$$

Number of paths of length 4 from a to $c = (1,3)^{th}$ entry in $A^4 = 4$.

★ <u>**Illustrative Examples**</u> ★

Example 2.4. *For the following graph G. Find*

   i) All paths from v_1 to v_8.   *ii) Circuits of lengths of 4 and 6.*

  iii) Trails of length 3 in G from v_3 to v_6.

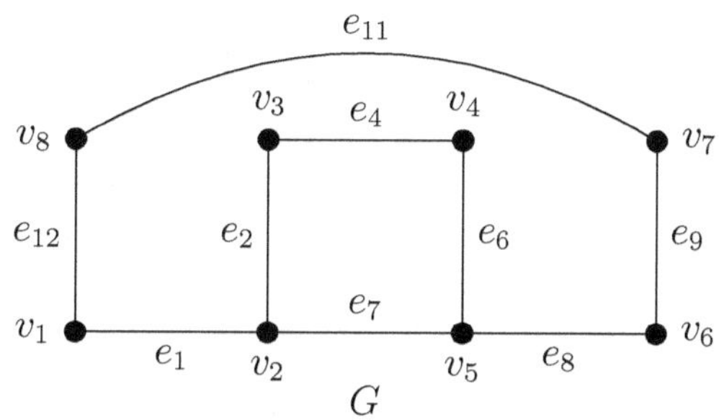

Figure 2.5

Solution:

  i) $P_1 : v_1 \ e_{12} \ v_8$

    $P_2 : v_1 \ e_1 \ v_2 \ e_7 \ v_5 \ e_8 \ v_6 \ e_9 \ v_7 \ e_{11} \ v_8$

    $P_3 : v_1 \ e_1 \ v_2 \ e_2 \ v_3 \ e_4 \ v_4 \ e_6 \ v_5 \ e_8 \ v_6 \ e_9 \ v_7 \ e_{11} \ v_8$

    These are the only possible paths from v_1 to v_8 in G.

  ii) $C_1 : v_2 \ e_2 \ v_3 \ e_4 \ v_4 \ e_6 \ v_5 \ e_7 \ v_2$ length $(C_1)=4$

    $C_2 : v_1 \ e_1 \ v_2 \ e_7 \ v_5 \ e_8 \ v_6 \ e_9 \ v_7 \ e_{11} \ v_8 \ e_{12} \ v_1$ length $(C_2)=6$

    These are the only possible circuits of lengths of 4 and 6.

  iii) $W_1 : v_3 \ e_4 \ v_4 \ e_6 \ v_5 \ e_8 \ v_6$ length $(W_1)=3$ $W_2 : v_3 \ e_2 \ v_2 \ e_7 \ v_5 \ e_8 \ v_6$ length $(W_2)=3$.

    These are the only possible trails of length 3 in G from v_3 to v_6.

Example 2.5. *Show that a connected graph G with n vertices has atleast $n-1$ edges.*

Solution: Consider the graph $G = (\{H_1, H_2, \cdots, H_n, \phi\})$. It has n components. When we add an edge between a pair of vertices in different components of G, then number of components is reduced by 1. So the addition of $n-1$ edges between suitable pairs of vertices makes G a connected graph. Hence a connected graph with n vertices has atleast $n-1$ edges.

Example 2.6. *Show that a simple graph G with n vertices and more than* $\dfrac{(n-1)(n-2)}{2}$ *edges is connected.*

Solution: We know that a simple graph with n vertices and k components has atmost $\dfrac{(n-k)(n-k+1)}{2}$ edges.

This means that in a simple disconnected graph G, the number of edges

$$e \leq \frac{(n-k)(n-k+1)}{2}$$

Equivalently if $e > \dfrac{(n-k)(n-k+1)}{2}$ then G is connected.

Putting $k = 1$ this imples that if $e > \dfrac{n(n-1)}{2}$ then graph G is connected. But, in this case, it will be a simple graph. Putting $k = 2$, we have, if $e > \dfrac{(n-2)(n-1)}{2}$ then G is connected and simple.

Example 2.7. *Let $V = \{0, 1, 2, 3, 4, 5, 6, 7, 8, 9, 10\}$ be the vertex set of a graph G. An edge between x and y exists when $x - y$ is divisible by 3. Draw a graph G. Explain why G is not connected. What the component of G represent.*

Solution: From the given condition of divisibility, we get following pairs of adjacent vertices.

$\{(0,0), (1,1), (2,2), (3,3), (4,4), (5,5), (6,6), (7,7), (8,8), (9,9), (10,10), (0,3), (1,4), (2,5), (3,6), (4,7),$ $(5,8), (6,9), (7,10), (0,6), (1,7), (2,8), (3,9), (4,10), (0,9), (1,10)\}$. The graph G is shown below.

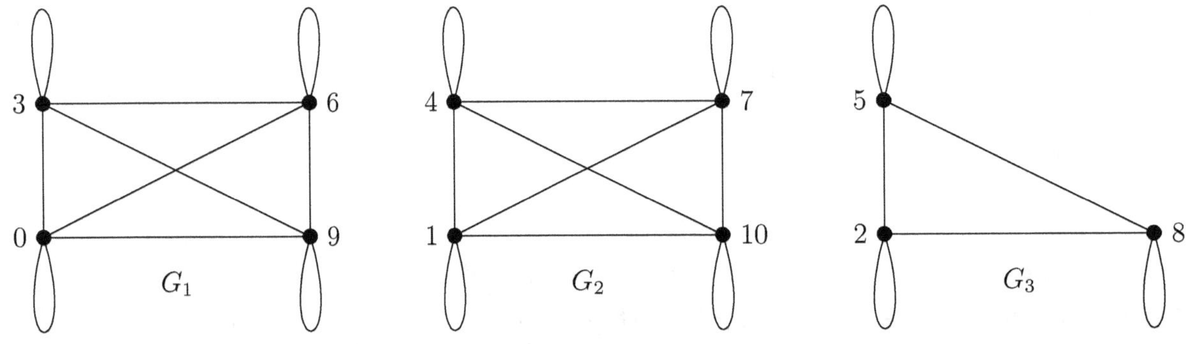

Figure 2.6

The graph G is not connected as it has three component G_1, G_2, G_3.

Components are equivalence classes with respect to equivalence relation congruence modulo 3.

Exercise: 2.1

1. Does each of these lists of vertices form a path in the following graph? Which paths are simple? Which are circuits? What are the lengths of those that are paths?

    i) a, e, b, c, b ii) a, e, a, d, b, c, a

iii) e, b, a, d, b, e iv) $c, b, d, a, e, c.$

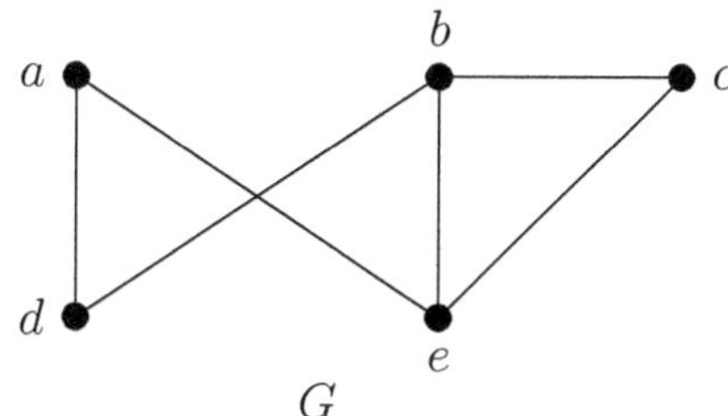

2. Find any two cycles in the following graphs.

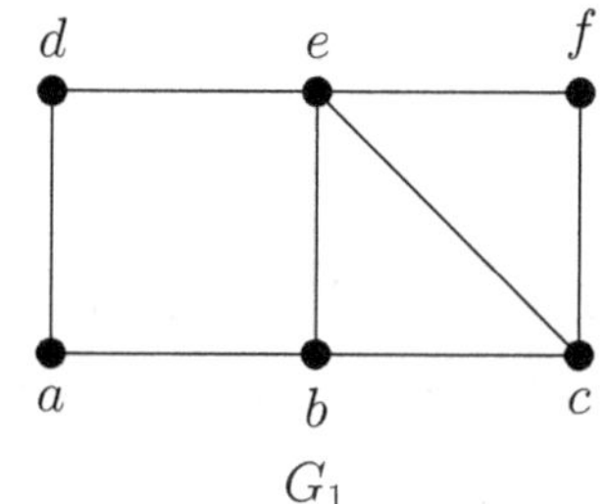 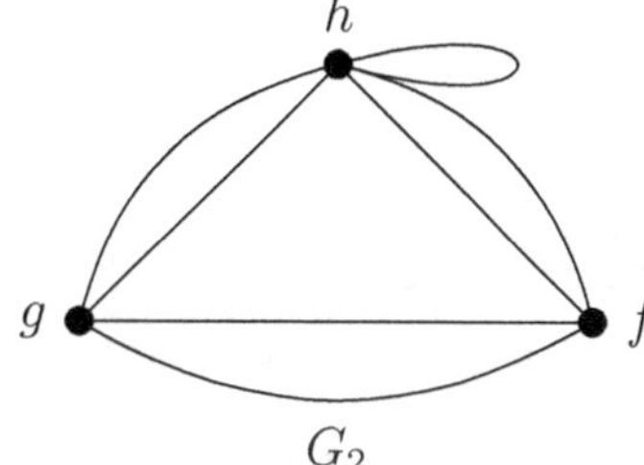

3. Show that if a and b are the only two odd degree vertices of a graph G, then a and b are connected in G.

4. Show that if G is a connected graph, then it is possible to remove vertices to disconnect G if and only if G is not a complete graph.

5. Let G be a disconnected graph with n vertices where n is even. If G has two components each of which is complete, prove that G has a minimum of $\dfrac{n(n-2)}{4}$ edges.

6. A simple graph G has vertex set $V = \{2, 3, 4, 6, 7, 8, 9, 10, 11, 12\}$. an edge exists between vertices x and y if $x \neq y$ and if x divides y or y divides x. Draw the graph G. Is G connected?

7. How many paths of length 4 are there from a to d in the following graph?

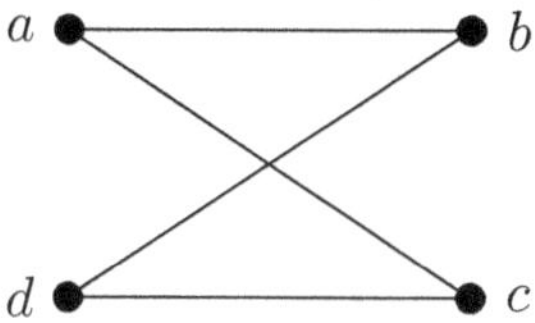

8. Find number of paths of length 3 between any two different vertices in K_4.

Solutions

1. i) Yes, it is a path. Not a simple path. Length $= 4$.

    ii) Not a path.

    iii) Not a path.

    iv) Yes it is a simple circuit of length 5.

2. Cycles in G_1 : $a - b - e - d - a$

$$b - e - c - b.$$

Cycles in G_2 : $g - h - f - g$

$$f - g - h - h - f.$$

3. If G is connected, nothing to prove. Let G be disconnected. If possible assume that a and b are not connected. Then a and b lie in the different components of G. Hence the component of G containing a (similarly containing b) contains only one odd degree vertex a, which is not possible as each component of G is itself a connected graph and in a graph number of odd degree vertices should be even. Therefore a and b lie in the same component of G. Hence they are connected.

5. Let x be the number of vertices in one of the components. Then the other component has nx number of vertices since both components are complete graphs, the number of edges they have are $\dfrac{x(x-1)}{2}$ and $\dfrac{(n-x)(n-x-1)}{2}$ respectively.

Therefore, the total number of edges in G is

$$m = \frac{x(x-1)}{2} + \frac{(n-x)(n-x-1)}{2} = x^2 - nx + \frac{n}{2}(n-1).$$

$$\Rightarrow \quad \frac{dm}{dx} = 2x - n, \quad \frac{d^2 m}{dx^2} = 2.$$

Therefore, m is minimum when $2xn = 0$. Hence $x = \dfrac{n}{2}$.

$$\text{Min } m = \left(\frac{n}{2}\right)^2 - n\left(\frac{n}{2}\right) + \frac{n}{2}(n-1)$$
$$= \frac{n(n-2)}{4}.$$

6.

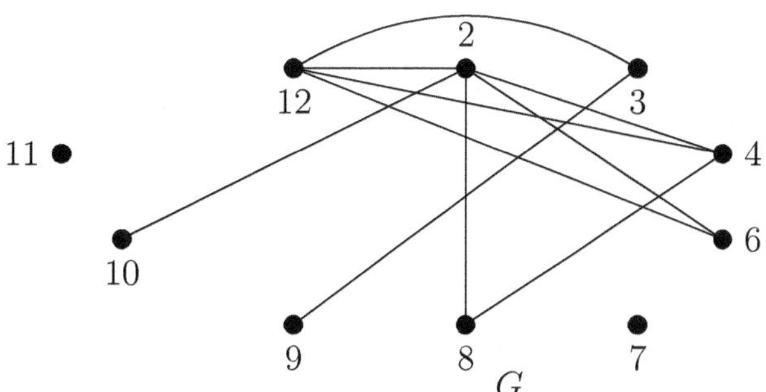

G is not connected.

7. 1.

8. 7.

2.4 Isthmus and Cut vertex

Definition 2.9 (Isthmus). *An edge e of a graph G is called an isthmus (or **bridge or cut edge**) if the number of components in $G - e$ is more than that in G.*

In otherwords, If e is an edge of G, such that $G - e$ is disconnected graph.

Consider the following graph.

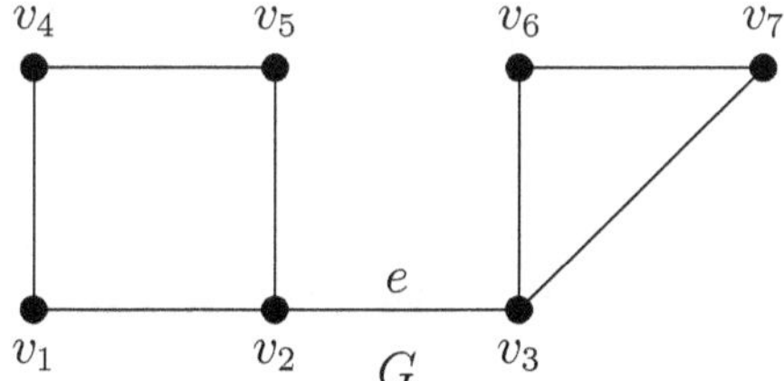 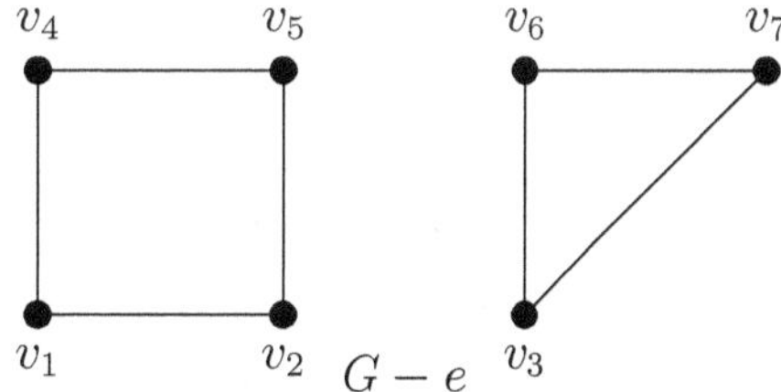

Figure 2.7

Note that after deletion of edge $'e'$ from G, G becomes disconnected. Hence, the edge e is an isthmus(bridge).

Definition 2.10 (Cut vertex). *A vertex $v \in V$ in a graph G is called a cut vertex if the number of components in $G - v$ is more than that in G.*

In otherwords, If u is a vertex of G, such that $G - u$ is disconnected graph then u is a cut vertex.

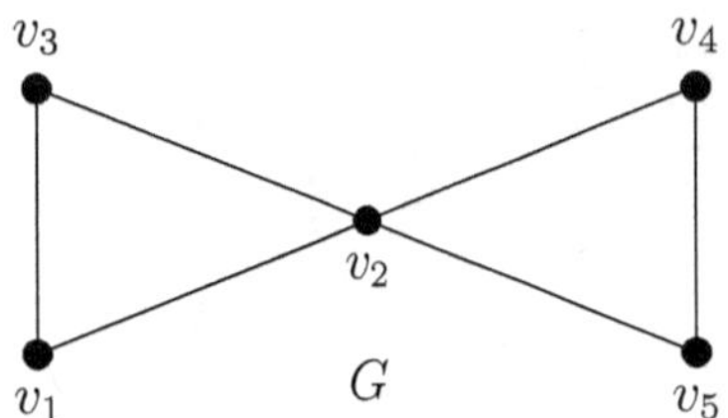
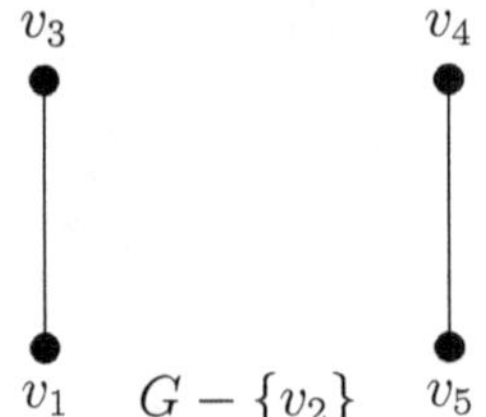

Figure 2.8

The vertex v_2 is cut vertex of G, because $G - \{v_2\}$ is disconnected.

Theorem 2.5. *Let $G(V, E)$ be any graph. The edge e of a graph G is an isthmus if and only if e does not belong to any circuit in G*

Proof: In the graph $G(V, E)$, let the edge e be an isthmus with u_1 and u_2 as its end vertices. Then by definition, $G - e$ consists of two components and the vertices u_1 and u_2 lie in these two components. We prove that e is not in any circuit of G. If possible, suppose e is in the circuit C of G.

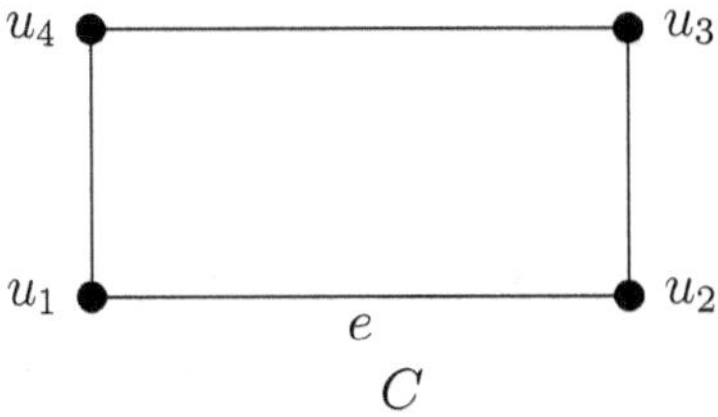

Now by removing e from G, the vertices u_1 and u_2 still remain connected through the path which is a part of circuit C. This contradicts to the definition of isthmus. Hence e cannot lie in any circuit of G.

Conversly, assume that the edge e does not lie in any circuit of the graph G. therefore if u_1 and u_2 are two vertices of e then $u_1 \, e \, u_2$ is the only path joining them. This implies that after removing the edge e from G, the vertices u_1 and u_2 are disconnected. Hence e is an isthmus.

2.4.1 Connectivity

Definition 2.11. *Let G be a connected graph. The minimum number of vertices whose removal from G results in a disconnected or trivial graph (single vertex) is called the **vertex connectivity** of G. The vertex connectivity of G is denoted by $\kappa(G)$.*

Remark 2.2.   *1. If $\kappa(G) = 1$ then, G has a vertex v such that $G - v$ is not connected and the vertex v is called a cut vertex.*

2. If $G = K_n$, the complete graph with n vertices then, $\kappa(G) = n - 1$.

4. If a graph G has a bridge then, the vertex connectivity of G, i.e., $\kappa(G) = 1$.

Consider the following graph.

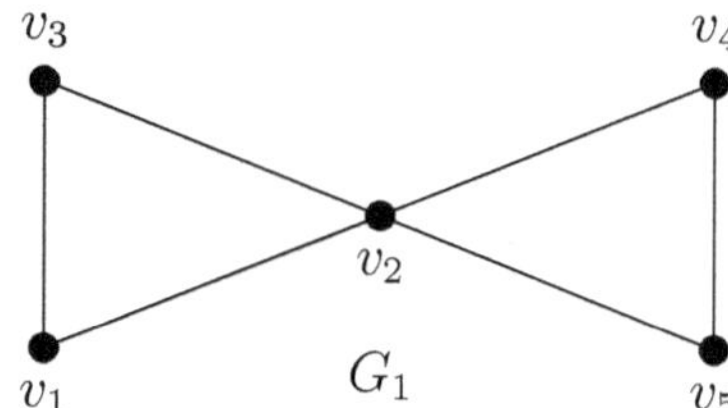 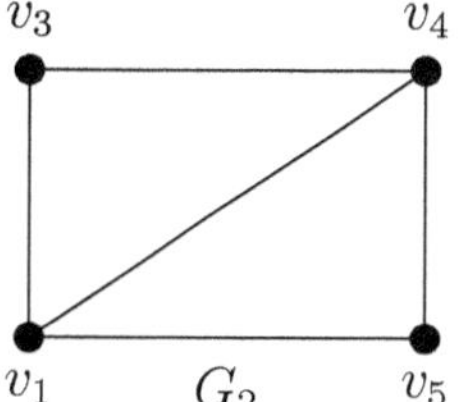

Figure 2.9

In the graph G_1, the vertex v_2 is a cut vertex and $\kappa(G_1) = 1$.

In the graph G_2, the deletion of any single vertex does not result in a disconnected graph. However, after deleting the two vertices v_1 and v_4 the remaining graph is disconnected. Therefore, $\kappa(G_2) = 2$.

Definition 2.12. *Let G be a connected graph. The **edge connectivity** of G is the minimum number of edges whose removal from G results in a disconnected or a trivial graph.*
The edge connectivity of G is denoted by $\lambda(G)$.

Remark 2.3. *1. If G is a disconnected graph, then $\lambda(G) = 0$.*

2. If G connected graph and G has a bridge, then the edge connectivity of G is one.

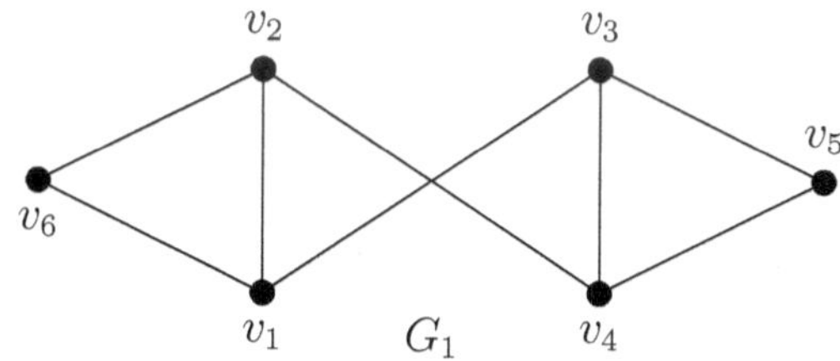 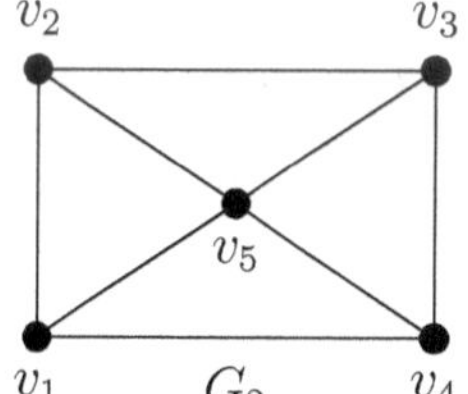

Figure 2.10

The edge connectivity in Graph G_1 is $\lambda(G_1) = 2$.
The edge connectivity in Graph G_2 is $\lambda(G_2) = 3$.

Theorem 2.6. *The edge connectivity of a connected graph G cannot exceed the minimum degree of G, i.e., $\lambda(G) \leq \delta(G)$. Where $\delta(G)$ denotes the smallest degree in a graph G.*

Proof: Let G be a connected graph and v be a vertex of minimum degree in G. Then the removal of edges incident with the vertex v disconnects the vertex v from the graph G. Thus the set of all edges incident with the vertex v forms a cut set of G. Hence, the edge connectivity of G cannot exceed the minimum degree of v, i.e., $\lambda(G) \leq \delta(G)$.

Theorem 2.7. *Let G be a connected graph with n vertices and e edges. Then $\lambda(G) \leq \left\lceil \dfrac{2e}{n} \right\rceil$, where*

Proof: Let $\delta(G)$ be the smallest degree of a vertex in G. Then the degree of each of n vertices is greater than or equal to $\delta(G)$.

$$\therefore \qquad \text{Total degree} \geq n\,\delta(G)$$

$$\therefore \qquad 2e \geq n\,\delta(G)$$

$$\Rightarrow \qquad \delta(G) \leq \frac{2e}{n}$$

$$\text{But we know that,} \qquad \lambda(G) \leq \delta(G)$$

$$\Rightarrow \qquad \lambda(G) \leq \frac{2e}{n}.$$

Now, since $\lambda(G)$ must be an integer, we have $\lambda(G) \leq \left\lceil \dfrac{2e}{n} \right\rceil$.

Theorem 2.8. *The vertex connectivity of a graph G is always less than or equal to the edge connectivity of G i.e.,* $\kappa(G) \leq \lambda(G)$.

Proof: If graph G is disconnected or trivial then $\kappa(G) = \lambda(G) = 0$. If G is connected and has a bridge e, then $\lambda(G) = 1$. In this case $\kappa(G) = 1$, since either G has a cut vertex incident with e.

$\therefore$ $\kappa(G) \leq \lambda(G)$ when $\lambda(G) = 0$ or 1, finally let us suppose that $\lambda(G) \geq 2$. The G has $\lambda(G)$ edges whose removal disconnects G. Clearly the $\lambda(G) - 1$ of these edges produces a graph with a bridge $e = \{u, v\}$. For each of these $\lambda(G) - 1$ edges select an incident vertex which is different from u or v. The removal of these points (vertices) also removes $\lambda(G) - 1$ edges and if the resulting graph is disconnected then $\kappa(G) \leq \lambda(G) - 1 \leq \lambda(G)$. If not the edge $e = \{u, v\}$. is a bridge and hence the removal of u and v will result in either a disconnected or a trivial graph. Hence $\kappa(G) \leq \lambda(G)$ in each case and this completes the proof of the theorem.

Remark 2.4. *Thus, the vertex connectivity of a graph does not exceed the edge connectivity and edge connectivity of a graph cannot exceed the minimum degree of G. Therefore we get the following relation*

$$\kappa(G) \leq \lambda(G) \leq \delta(G) \leq \left\lceil \frac{2e}{n} \right\rceil$$

★ <u>Illustrative Examples</u> ★

Example 2.8. *Find all bridges and cut vertices of the following graph*

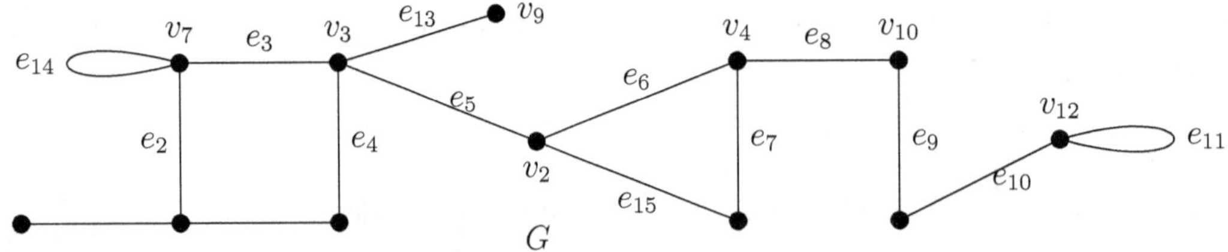

Solution: The bridges (isthmus) in a graph are precisely those edges which does not belong to any circuit of the given graph. In the above graph, the the edges e_5, e_8, e_9, e_{10}, e_{12}, e_{13} do not belongs to any circuit. Therefore they are bridges.

The cut vertices in a graph are precisely those vertices such that every path joining any two vertices x and y in a graph G passes through v. In the above graph the vertices $\{v_2,\ v_3,\ v_4,\ v_6,\ v_{10},\ v_{11}\}$ are cut vertices.

Example 2.9. *Find the edge connectivity and the vertex connectivity of the graphs in Fig. 2.11.*

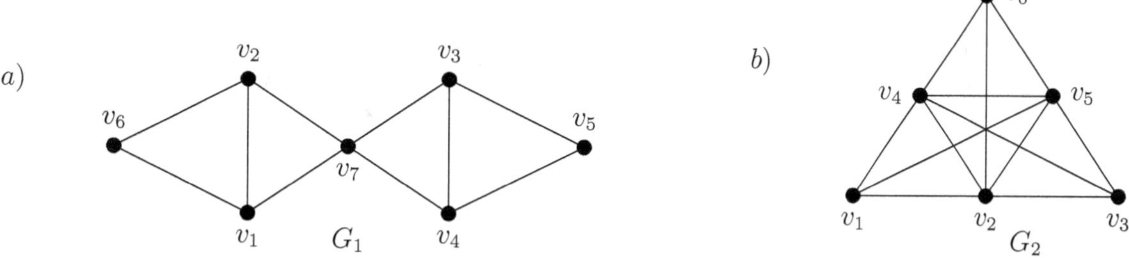

Figure 2.11

Solution:

a) In graph G_1, v_7 is a cut vertex. Therefore $\kappa(G) = 1$. Smallest degree of vertex is 2. That is $\delta(G) = 2$. Therefore, by using

$$\kappa(G) \leq \lambda(G) \leq \delta(G)$$

$$1 \leq \lambda(G) \leq 2$$

But G has no isthmus, hence $\lambda(G) = 2$.

b) The minimum number of edges whose removal disconnects the graph G_2 is 3 and the minimum number of vertices required to disconnect the graph is 3.

∴   Edge connectivity $\lambda(G) = 3$

Vertex connectivity $\kappa(G) = 3$..

Example 2.10. *Is it possible to construct a graph on 7 vertices and 18 edges with edge connectivity 5? Justify.*

Solution: Yes it is possible to construct such graph. As we have given $n = 7$, $e = 18$, $\lambda(G) = 5$.

$$\text{We know that} \qquad \lambda(G) \leq \left\lceil \frac{2e}{n} \right\rceil$$

$$\therefore \qquad 5 \leq \left\lceil \frac{2 \times 18}{7} \right\rceil$$

$$\Rightarrow \qquad 5 \leq \lceil 5.14 \rceil = 5$$

Example 2.11. *Construct the graph G for each of the following conditions.*

a) $\kappa(G) = \lambda(G) = \delta(G) = 2$

b) $\kappa(G) = 1,\ \lambda(G) = \delta(G) = 2$

c) $\kappa(G) = 2,\ \lambda(G) = \delta(G) = 3$

d) $\kappa(G) = 1,\ \lambda(G) = 2,\ \delta(G) = 3.$

Solution:

a) $\kappa(G) = \lambda(G) = \delta(G) = 2$ b) $\kappa(G) = 1,\ \lambda(G) = \delta(G) = 2$

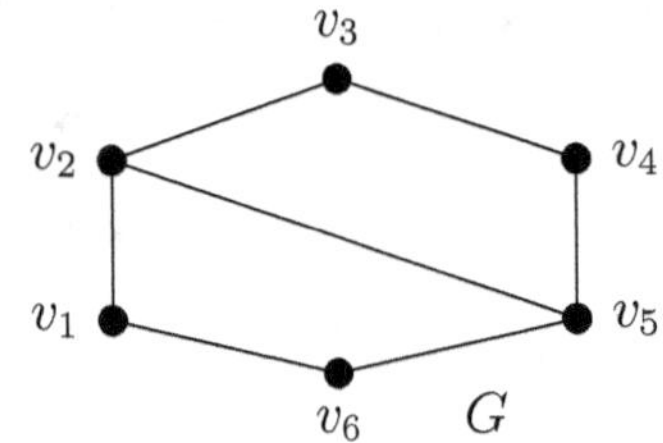
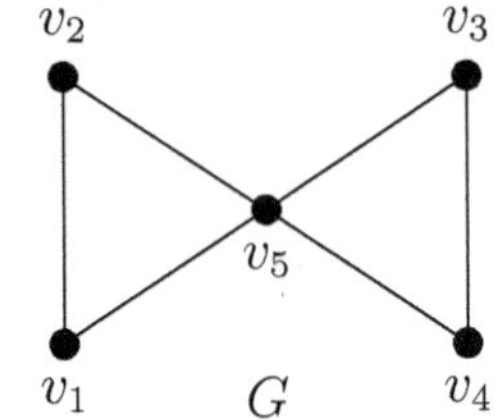

c) $\kappa(G) = 2,\ \lambda(G) = \delta(G) = 3$ d) $\kappa(G) = 1,\ \lambda(G) = 2,\ \delta(G) = 3.$

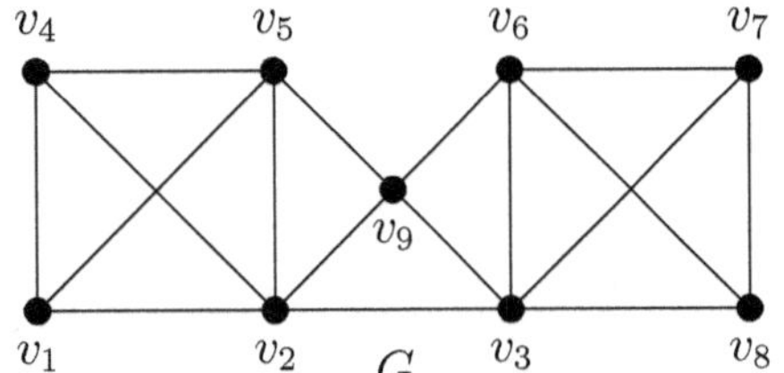
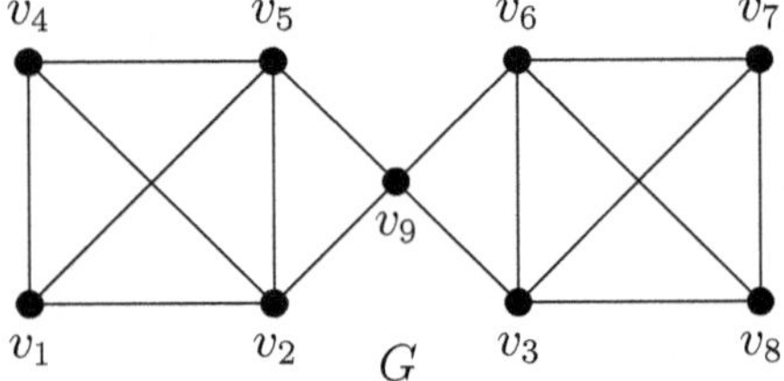

Figure 2.12

Exercise: 2.2

1. State true or false with justification.

   i) Every subgraph of a connected graph is connected.

   ii) If G has bridge then it has a cut vertex.

   iii) Vertex connectivity of a cycle is 3.

   iv) A bipartite graph is always connected.

   v) There exists a graph with vertex connectivity 3 and edge connectivity 2.

2. Draw graph $K_{4,3}$ and find its vertex connectivity and edge connectivity.

3. Give an example of a connected graph G such that $\kappa(G) < \lambda(G) < \delta(G)$.

4. Find all bridges and cut vertices of the following graph.

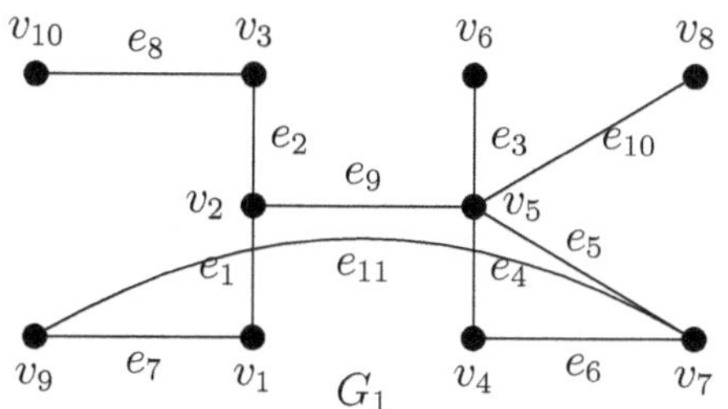 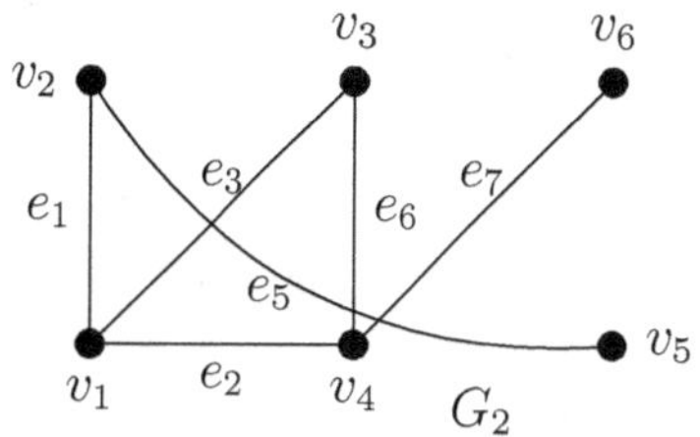

5. For each of these graphs, find $\kappa(G)$, $\lambda(G)$ and $\delta(G)$ also determine which of the two inequalities in $\kappa(G) \le \lambda(G) \le \delta(G)$. are strict.

$i)$ 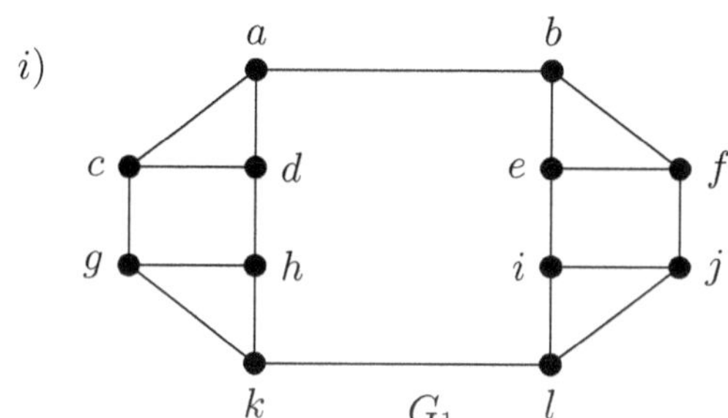 $ii)$ 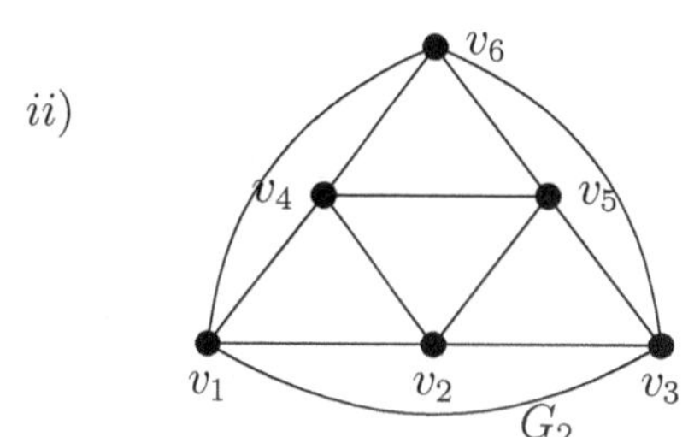

6. Show that a simple graph G is bipartite if and only if it has no circuits with an odd number of edges.

Solutions

1. i) False.

   ii) True.

   iii) False.

   iv) False.

   v) False.

2.

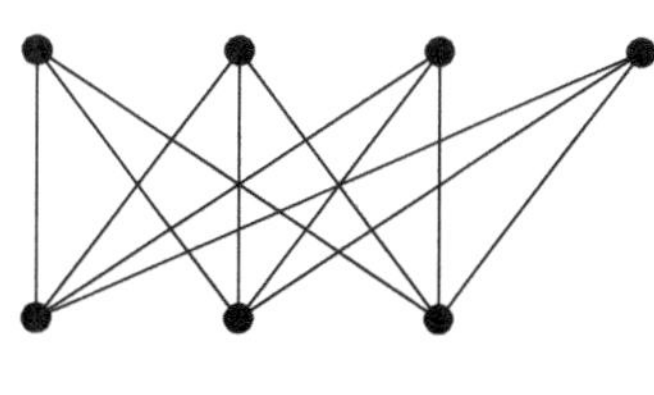

$$K_{4,3}$$

3.

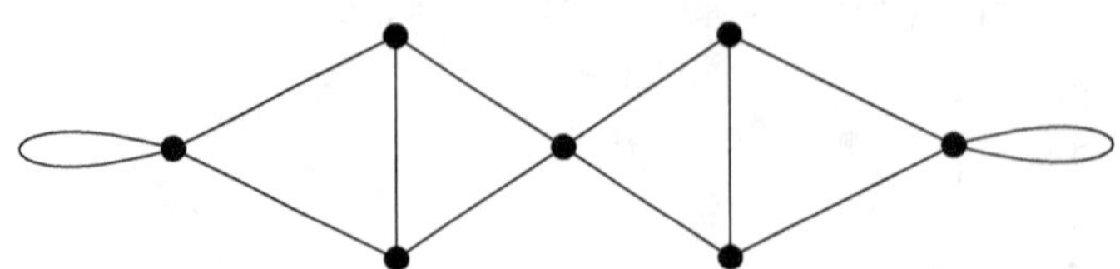

$$\kappa(G) = 1,\ \lambda(G) = 2,\ \delta(G) = 3.$$

3. G_1 : bridges $\qquad e_8, e_2, e_3, e_{10}$

$\qquad$ cutvertices $\qquad v_3, v_5, v_2.$

$\quad G_2$: bridges $\qquad e_7, e_5$

$\qquad$ cutvertices $\qquad v_2, v_4, v_1.$

5. i) $\kappa(G) = 2,\ \lambda(G) = 2,\ \delta(G) = 3.$

$\quad$ ii) $\kappa(G) = 4,\ \lambda(G) = 4,\ \delta(G) = 4.$

2.5 A Shortest-Path algorithm

Definition 2.13. *A graph in which every edge is assigned the weight (a non-negative real number) is called a* **weighted graph.** *Weight of an edge e is denoted by $w(e)$.*

Weight of a graph is denoted by $W(G)$ and it is defined as the sum of weights of all edges in G. Consider the following graph.

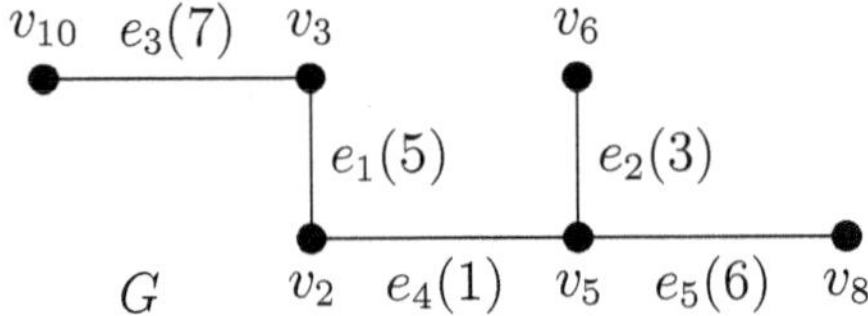

Weight of G:

$$w(G) = w(e_1) + w(e_2) + w(e_3) + w(e_4) + w(e_5)$$
$$= 5 + 3 + 7 + 1 + 6 = 22.$$

The shortest path problem is concerned with finding the least cost (that costs minimum) path from an originating node in a weighted graph to a destination node in that graph. There are several different algorithms that find a shortest path between two vertices in a weighted graph. One such algorithm is developed by Dijkstra in the early 1960s for finding the shortest path in a graph with non-negative weight associated with edge without explicitly enumerating all possible paths. This algorithm is based on technique of finding the shortest path from the vertex a to the vertex b of a graph G, the labelling technique is used in **Dijkstra's Algorithm.**

- ## Dijkstra's Algorithm

The algorithm relies on a series of iterations. A distinguished set of vertices is constructed by adding one vertex at each iteration. A labeling procedure is carried out at each iteration. In this labeling procedure, a vertex w is labeled with the length of a shortest path from a to w that contains only vertices already in the distinguished set. The vertex added to the distinguished set is one with a minimal label among those vertices not already in the set.

Let G has vertices $a = v_0, v_1, \cdots, v_n = z$ and lengths $w(v_i, v_j)$ where $w(v_i, v_j) = \infty$ if $\{v_i, v_j\}$ is not an edge in G. It begins by labeling a with 0 and the other vertices with ∞. We use the notation $\lambda(a) = 0$ and $\lambda(v) = \infty$ for these labels before any iterations have taken place. These labels are the lengths of shortest paths from a to the vertices, where the paths contain only the vertex a. We denote set of vertices having temporary label by T. If $v \in T$ is not adjacent to 'a' then $\lambda(v)$ will not be changed. If $v \in T$ is adjacent to 'a' then $\lambda(v)$ will be changed to

$$\min\{\lambda(v), \lambda(a) + \text{weight of the edge}(a, v)\}$$

This procedure is iterated by successively adding vertices to the distinguished set until z is added. When z is added to the distinguished set, its label is the length of a shortest path from a to z.

We summarize the above process to find the weight of the shortest path from the vertex 'a' to the vertex 'z'.

1. Consider a connected graph G having vertices $a = v_0, v_1, \cdots, v_n = z$

2. Define a set T of vertices and initialize it to empty set. As the algorithm progresses, the set T will store those vertices to which a shortest path has been found.

3. Initially take $\lambda(a) = 0$ and $\lambda(v) = \infty$ for $v \neq a$; assign $T = v$.

4. Select the vertex v of minimum λ value and assign permanent label to v.

5. if $v = z$ then stop.

6. If $v \neq z$ then update the labels of vertices in $T - \{v\}$.

7. Change T to $T - \{v\}$.

8. Go to step 4.

★ <u>**Illustrative Examples**</u> ★

Example 2.12. *Apply Dijkstras algorithm to the graph given below and find the shortest path from a to z.*

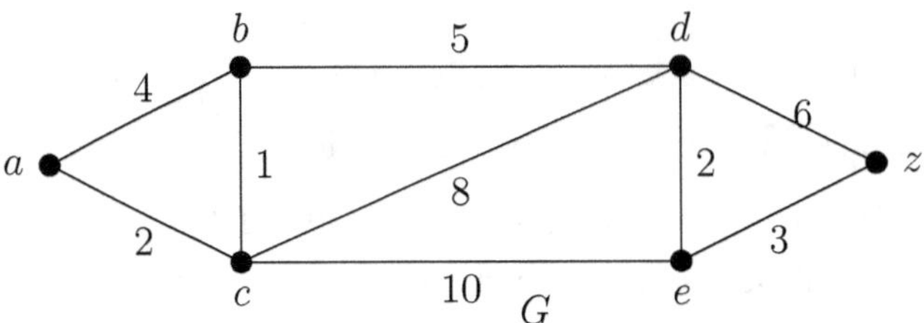

Figure 2.13

Solution: The initial labelling is given by

Vertex V	**a**	**b**	**c**	**d**	**e**	**z**
$\lambda(v)$	0	∞	∞	∞	∞	∞
T	a	b	c	d	e	z

Iteration 1 :

$\lambda(a) = 0$ is minimum. T becomes $T - \{a\}$. The vertices in T adjacent to a are b and c.

$\lambda(b) = \min \{ \text{ old } \lambda(b), \lambda(a) + w(a,b)\},$
$\qquad = \min \{\infty, 0 + 4\} = 4$

$\lambda(c) = \min \{ \text{ old } \lambda(c), \lambda(a) + w(a,c)\}$
$\qquad = \min \{\infty, 0 + 2\} = 2$

Vertex V	**a**	**b**	**c**	**d**	**e**	**z**
$\lambda(v)$	0	4	2	∞	∞	∞
T	-	b	c	d	e	z

Hence minimum label is $\lambda(c) = 2$.

Iteration 2 :

$\lambda(c) = 2$ is minimum. T becomes $T - \{a, c\}$. The vertices in T adjacent to c are b, d and e.

$\lambda(b) = \min \{ \text{ old } \lambda(b), \lambda(c) + w(c,b)\},$
$\qquad = \min \{4, 2 + 1\} = 3$

$\lambda(d) = \min \{ \text{ old } \lambda(d), \lambda(c) + w(c,d)\}$
$\qquad = \min \{\infty, 2 + 8\} = 10$

$\lambda(e) = \min \{ \text{ old } \lambda(e), \lambda(c) + w(c,e)\}$
$\qquad = \min \{\infty, 2 + 10\} = 12$

Vertex V	a	b	c	d	e	z
$\lambda(v)$	0	3	2	10	12	∞
T	-	b	-	d	e	z

Hence minimum label is $\lambda(b) = 3$.

Iteration 3 :

$\lambda(b) = 3$ is minimum. T becomes $T - \{a, b, c\}$. The vertices in T adjacent to b is d.

$$\lambda(d) = \min \{ \text{ old } \lambda(d), \lambda(b) + w(b, d)\}$$
$$= \min \{10, 3 + 5\} = 8$$

Vertex V	a	b	c	d	e	z
$\lambda(v)$	0	3	2	8	12	∞
T	-	-	-	d	e	z

Hence minimum label is $\lambda(d) = 8$.

Iteration 4 :

$\lambda(d) = 8$ is minimum. T becomes $T - \{a, b, c, d\}$. The vertices in T adjacent to d are e and z.

$$\lambda(e) = \min \{ \text{ old } \lambda(e), \lambda(d) + w(d, e)\}, \qquad \lambda(z) = \min \{ \text{ old } \lambda(z), \lambda(d) + w(d, z)\}$$
$$= \min \{12, 8 + 2\} = 10 \qquad\qquad\qquad = \min \{\infty, 8 + 6\} = 14$$

Vertex V	a	b	c	d	e	z
$\lambda(v)$	0	3	2	8	10	14
T	-	-	-	-	e	z

Hence minimum label is $\lambda(e) = 10$.

Iteration 5:

$\lambda(e) = 10$ is minimum. T becomes $T - \{a, b, c, d, e\}$. The vertices in T adjacent to e is z.

$$\lambda(z) = \min \{ \text{ old } \lambda(z), \lambda(e) + w(e, z)\}$$
$$= \min \{14, 10 + 3\} = 13$$

Vertex V	a	b	c	d	e	z
$\lambda(v)$	0	3	2	8	10	13
T	-	-	-	-	-	z

Since only one vertex z is remaining and $\lambda(z) = 13$. Iteration stops. Thus the shortest distance between a and z is 13, and moreover, the shortest paths is $\{a, c, b, d, e, z\}$.

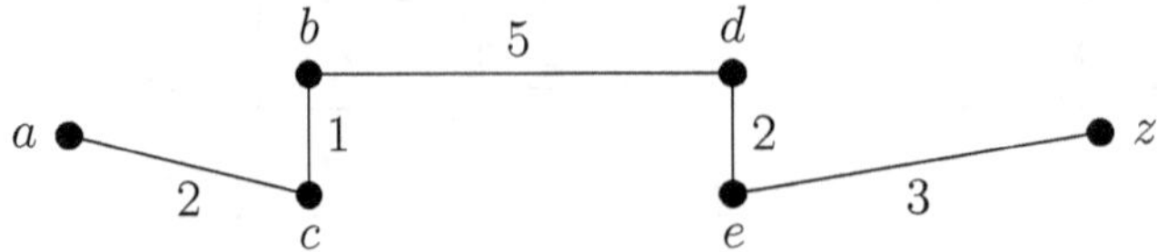

Figure 2.14

Example 2.13. *Find the length of a shortest path between a and z in the given weighted graph using Dijkstras algorithm.*

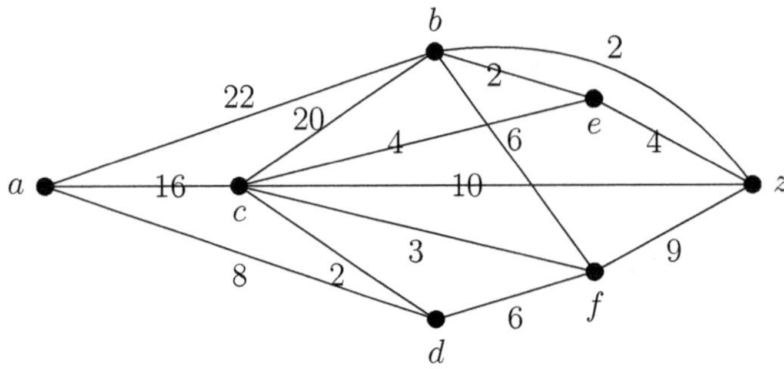

Figure 2.15

Solution: The initial labelling is given by

Vertex V	a	b	c	d	e	f	z
$\lambda(v)$	0	∞	∞	∞	∞	∞	∞
T	a	b	c	d	e	f	z

Iteration 1 :

$\lambda(a) = 0$ is minimum. T becomes $T - \{a\}$. The vertices in T adjacent to a are b c and d.

$\lambda(b) = \min \{ \text{ old } \lambda(b), \lambda(a) + w(a, b)\},$

$\quad = \min \{\infty, 0 + 22\} = 22$

$\lambda(c) = \min \{ \text{ old } \lambda(c), \lambda(a) + w(a, c)\}$

$\quad = \min \{\infty, 0 + 16\} = 16$

$\lambda(d) = \min \{ \text{ old } \lambda(d), \lambda(a) + w(a, d)\}$

$\quad = \min \{\infty, 0 + 8\} = 8$

Vertex V	a	b	c	d	e	f	z
$\lambda(v)$	0	22	16	8	∞	∞	∞
T		b		d		f	

Hence minimum label is $\lambda(d) = 8$.

Iteration 2:

$\lambda(d) = 8$ is minimum. T becomes $T - \{a, d\}$. The vertices in T adjacent to d are c and f.

$\lambda(c) = \min \{ \text{ old } \lambda(c), \lambda(d) + w(d, c)\},$

$\qquad = \min \{16, 8 + 2\} = 10$

$\lambda(f) = \min \{ \text{ old } \lambda(f), \lambda(d) + w(d, f)\}$

$\qquad = \min \{\infty, 8 + 6\} = 14$

Vertex V	**a**	**b**	**c**	**d**	**e**	**f**	**z**
$\lambda(v)$	0	22	10	8	∞	14	∞
T	-	b	c	-	e	f	z

Hence minimum label is $\lambda(c) = 10$.

Iteration 3:

$\lambda(c) = 10$ is minimum. T becomes $T - \{a, c, d\}$. The vertices in T adjacent to c are b, e, z and f.

$\lambda(b) = \min \{ \text{ old } \lambda(b), \lambda(c) + w(c, b)\},$

$\qquad = \min \{22, 10 + 20\} = 22$

$\lambda(e) = \min \{ \text{ old } \lambda(e), \lambda(c) + w(c, e)\}$

$\qquad = \min \{\infty, 10 + 4\} = 14.$

$\lambda(z) = \min \{ \text{ old } \lambda(z), \lambda(c) + w(c, z)\},$

$\qquad = \min \{\infty, 10 + 10\} = 20$

$\lambda(f) = \min \{ \text{ old } \lambda(f), \lambda(c) + w(c, f)\}$

$\qquad = \min \{14, 10 + 3\} = 13$

Vertex V	**a**	**b**	**c**	**d**	**e**	**f**	**z**
$\lambda(v)$	0	22	10	8	14	13	20
T	-	b	-	-	e	f	z

Hence minimum label is $\lambda(f) = 13$.

Iteration 4:

$\lambda(f) = 13$ is minimum. T becomes $T - \{a, c, d, f\}$. The vertices in T adjacent to f are b and z.

$\lambda(b) = \min \{ \text{ old } \lambda(b), \lambda(f) + w(f, b)\},$

$\qquad = \min \{22, 13 + 6\} = 19$

$\lambda(z) = \min \{ \text{ old } \lambda(z), \lambda(f) + w(f, z)\}$

$\qquad = \min \{20, 13 + 9\} = 20.$

Vertex V	**a**	**b**	**c**	**d**	**e**	**f**	**z**
$\lambda(v)$	0	19	10	8	14	13	20

Hence minimum label is $\lambda(e) = 14$.

Iteration 5:

$\lambda(e) = 14$ is minimum. T becomes $T - \{a, c, d, e, f\}$. The vertices in T adjacent to e are b and z.

$$\lambda(b) = \min \{ \text{ old } \lambda(b), \lambda(e) + w(e, b)\}, \qquad \lambda(z) = \min \{ \text{ old } \lambda(z), \lambda(e) + w(e, z)\}$$
$$= \min \{19, 14 + 2\} = 16 \qquad\qquad = \min \{20, 14 + 4\} = 18.$$

Vertex V	a	b	c	d	e	f	z
$\lambda(v)$	0	16	10	8	14	13	18
T	-	b	-	-	-	-	z

Hence minimum label is $\lambda(b) = 16$.

Iteration 6:

$\lambda(b) = 16$ is minimum. T becomes $T - \{a, b, c, d, e, f\}$. The vertices in T adjacent to b is z.

$$\lambda(z) = \min \{ \text{ old } \lambda(z), \lambda(b) + w(b, z)\}$$
$$= \min \{18, 16 + 2\} = 18$$

Vertex V	a	b	c	d	e	f	z
$\lambda(v)$	0	16	10	8	14	13	18
T	-	-	-	-	-	-	z

Hence minimum label is $\lambda(z) = 18$.

Since only one vertex z is remaining and $\lambda(z) = 13$. Iteration stops. Thus the shortest distance between a and z is 13 and the shortest paths is $\{a, d, c, e, z\}$.

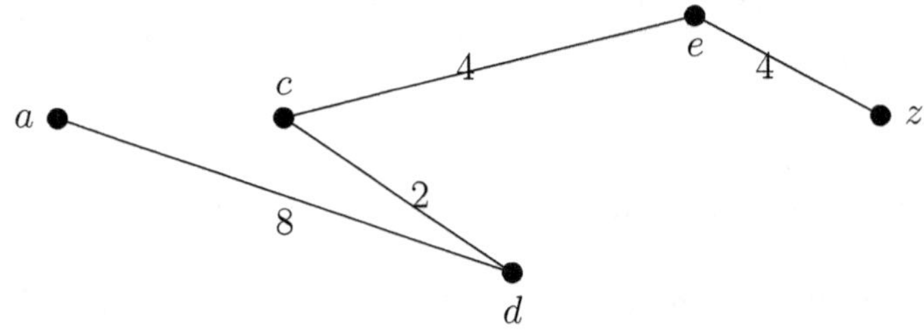

Exercise: 2.3

1. Find the shortest path from a to z in the graph below.

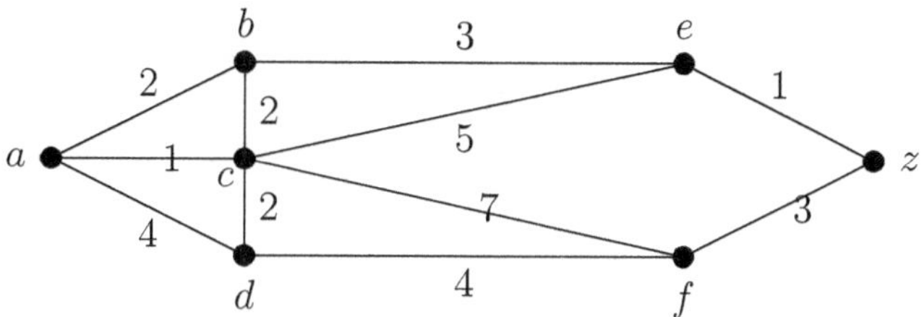

Figure 2.16

2. Find the shortest path from a to all other vertices of the following graph.

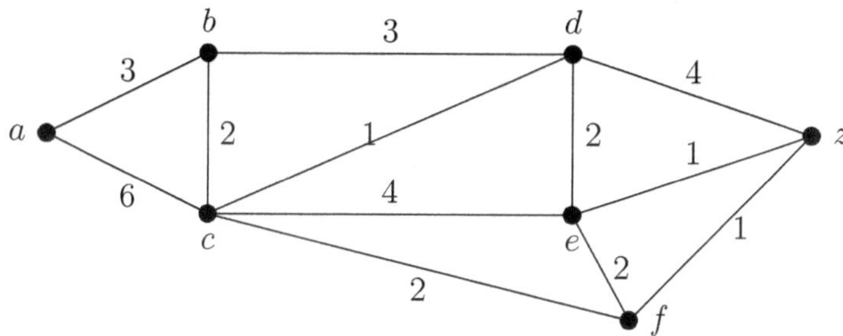

Figure 2.17

3. Find the shortest path from a to t in the graph below.

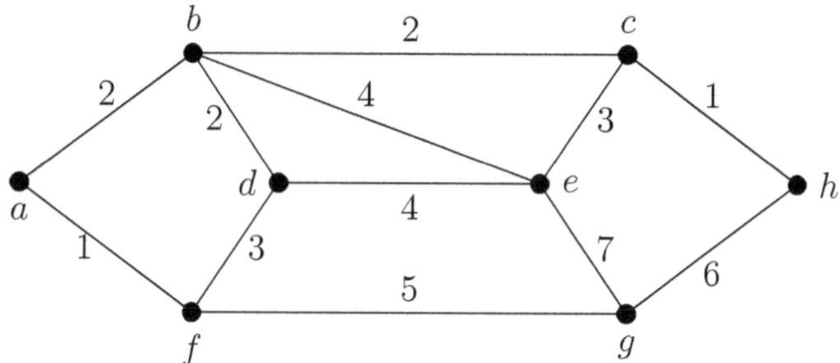

Figure 2.18

Solutions

1.

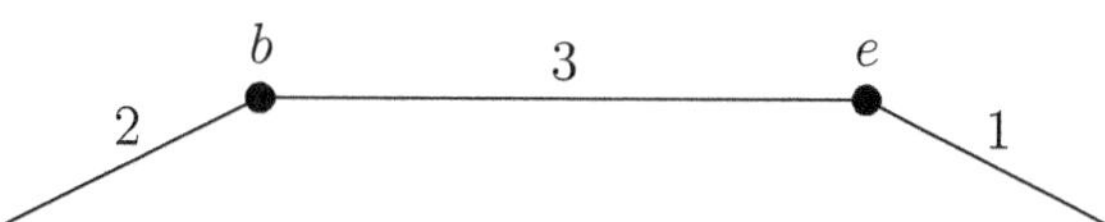

2.

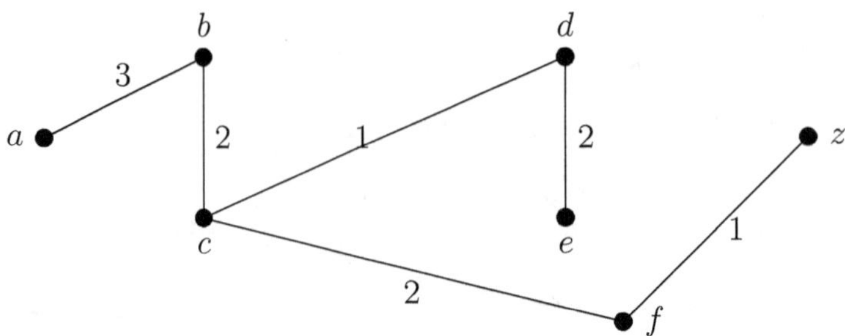

3.

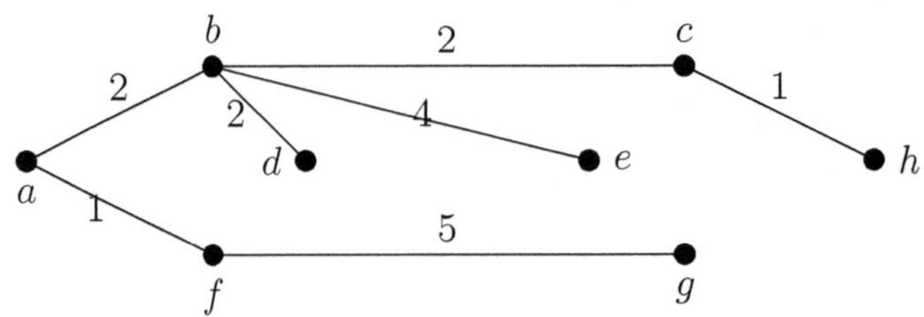

◆◆◆

Chapter 3

Euler and Hamilton Path

Introduction: The Swiss mathematician Leonhard Euler made the first use of Graph theory in 1736 to solve the well known Konigsberg problem. When this problem is represented using graphs, the problem reduced to the following question: 'Can we travel along the edges of a graph starting at a vertex and returning to it by traversing each edge of the graph exactly once?' A similar other question is ' Can we travel along the edges of a graph starting at a vertex and returning to it while visiting each vertex of the graph exactly once?' In this chapter, we will study these questions which give rise to Euler and Hamilton circuits. The chapter concludes with their application to Travelling salesman and Chinese Postman problem.

3.1   Euler Path and Euler circuit

Definition 3.1. *An **Euler circuit** in a graph G is a simple circuit containing every edge of G.*

Definition 3.2. *An **Euler path** in a graph G is a simple path containing every edge of G.*

Definition 3.3. *A graph G which admits an Euler circuit is called an Eulerian graph.*

In other words, if we can travel along the edges of the graph starting at a vertex and returning to it by traversing each edge of the graph exactly once then the graph is called an Eulerian graph.

Illustrative example: Consider the graph G_1 and G_2 given below :

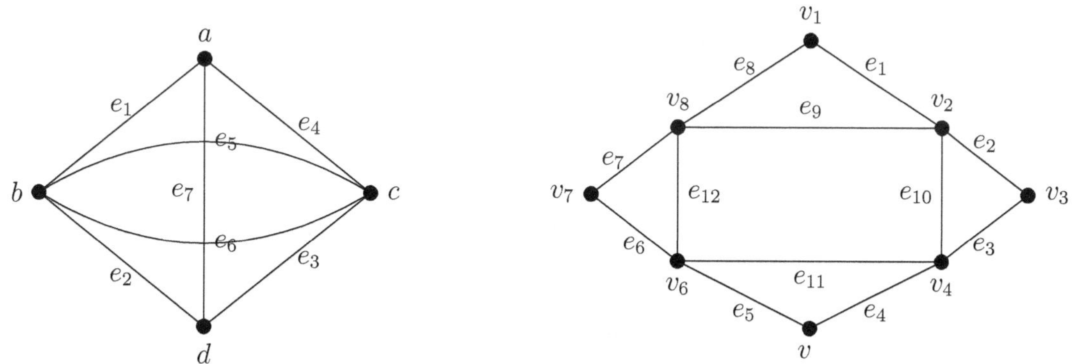

In graph G1, the path $a\ e_1 b\ e_2 d\ e_3 c\ e_6 b\ e_5 c\ e_4 a\ e_7 d$ is an Euler path, since it covers every edge.

In graph G2, the circuit $v_1 e_1 v_2 e_2 v_3 e_3 v_4 e_4 v_5 e_5 v_6 e_6 v_7 e_7 v_8 e_{12} v_6 e_{11} v_4 e_{10} v_2 e_9 v_8 e_8 v_1$ is the Euler circuit. Hence the graph G2 is an Eulerian graph.

In the following theorem we give the characteruzation of the Eulerian graph.

Theorem 3.1. *A connected graph G is Eulerian if and only if the degree of every vertex is even.*

Proof. Let G be a connected graph which is Eulerian. So it contains an Euler circuit. Suppose the Euler circuit begins and ends at vertex 'a'. The circuit contributes 1 to the degree of 'a' when it begins and ends at 'a'. Also it contributes 2 to the degree of 'a' each time it passes through 'a'. So degree of 'a' must be even. Now if 'b' is any vertex different from 'a', then 'b' is entered as many times it is left. Thus if Euler circuit visits 'b' r times then degree of 'b' must be 2r which is even. Therefore each vertex of G has even degree.

Conversely suppose that each vertex has even degree. So we can exit from each vertex we enter. We start with arbitrary vertex 'u' of G and by tracing edges of G complete the circuit $C_1 :\ u - u_1 - u_2 - \cdots - u.$

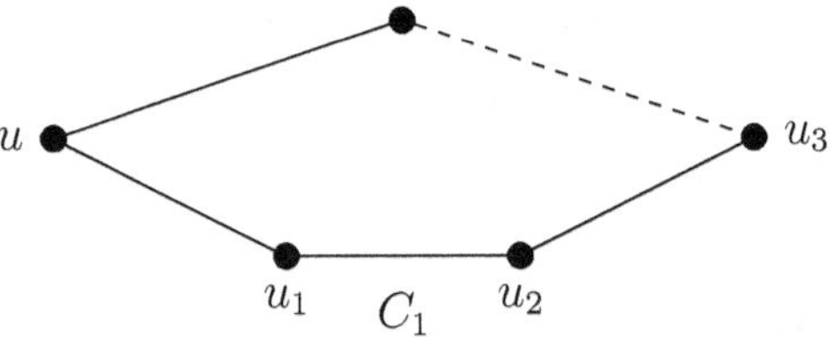

If this circuit C_1 contains all edges of G then C_1 is the required Euler circuit. If not, consider a graph $G_1 = G - C_1$. Again degree of each vertex in G_1 is even. Since G is connected, there is a vertex say u_i common in G and G_1. Now as before we trace the edges in G_1 and complete a circuit $C_2 : u_i - v_1 - v_2 - \cdots - u_i.$

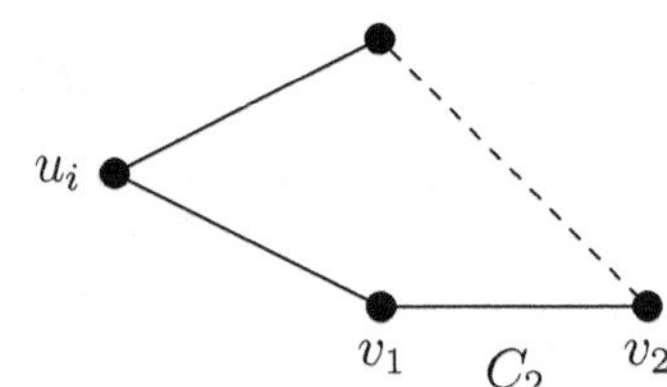

We can combine the above two circuits to get a circuit containing all covered edges as follows.

$$u - u_1 - u_2 - \cdots u_i - v_1 - v_2 \cdots u_i - u_{i+1} - \cdots u$$

If all the edges are covered in this then it gives the Euler circuit. Otherwise we continue the process with the graph $G_2 = G_1 - C_2$. We continue this till all the edges in G are covered and we have the Euler circuit. $\qquad \square$

Theorem 3.2. *A connected graph G has an Euler path if and only if it has exactly two vertices of odd degree.*

Proof. Suppose a connected graph G has an Euler path that begins at vertex 'a' and ends at vertex 'b'. The Euler path contributes 1 to the degree of 'a' when it begins and 2 to the degree of 'a' every time it visits 'a'. So degree of 'a' must be odd. Similarly the Euler path contributes 2 to the degree of b whenever it visits it and 1 when it ends at 'b'. So degree of 'b' is odd. Now if 'v' is any vertex different from 'a' and 'b' then the Euler circuit contributes 2 to the degree of 'v' every time it visits. Therefore degree of 'v' must be even. Hence G has exactly two vertices 'a' and 'b' of odd degree. Conversely if G has exactly two vertices say 'a' and 'b' of odd degree, then add a temporary edge between 'a' and 'b'. Now degree of every vertex in G is even. So by previous theorem, G contains an Euler circuit containing all edges of G. The removal of the temporary edge between 'a' and 'b' gives the Euler path that begins at 'a' and ends at 'b'. $\qquad \square$

Illustrative Example: Consider the graphs given below.

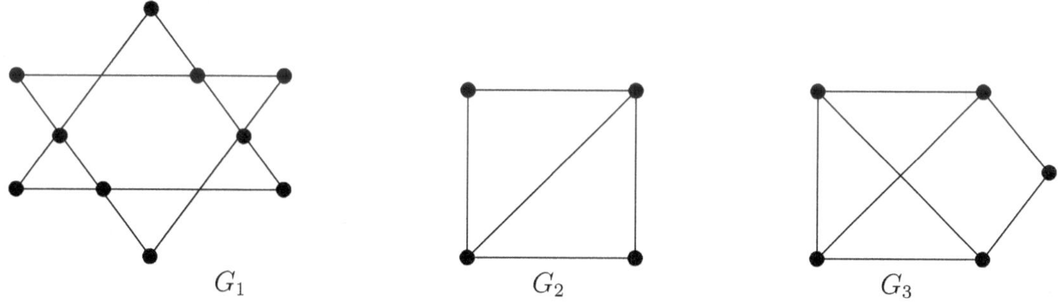

Figure 3.1

Note that degree of every vertex in G_1 is even. Hence G_1 contains an Euler circuit. In graph G_2, there are exactly two vertices of odd degree. Hence G_2 contains an Euler path. Graph G_3 does not contain an Euler circuit or an Euler path.

Konigsberg Problem : The city of Konigsberg Prussia (now called Kaliningrad and part of Russian republic) was divided into four sections by the branches of the Pregel river. These four sections included the two regions on the banks of the Pregel, Kneiph of Island and the region between the two branches of Pregel. These regions were linked by seven bridges as shown in the figure. 3.2.

The problem was to start at any of the four land areas of the city A,B,C, or D, walk over each of the seven bridges exactly once and return to the starting point.

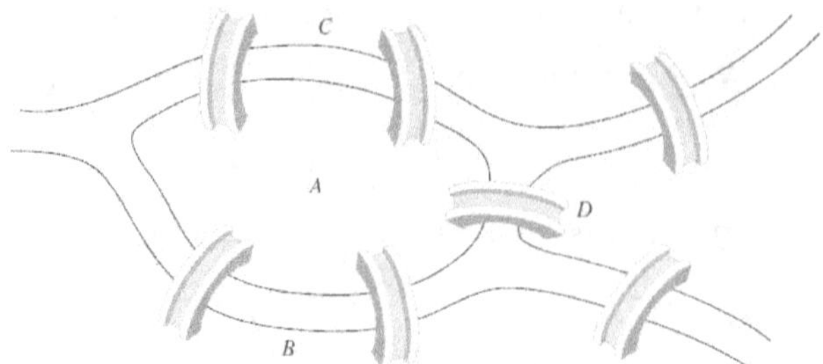

Figure 3.2

Euler represented this problem using graph. He used vertices to represent the land areas and edges to represent the bridges as shown in the following figure.

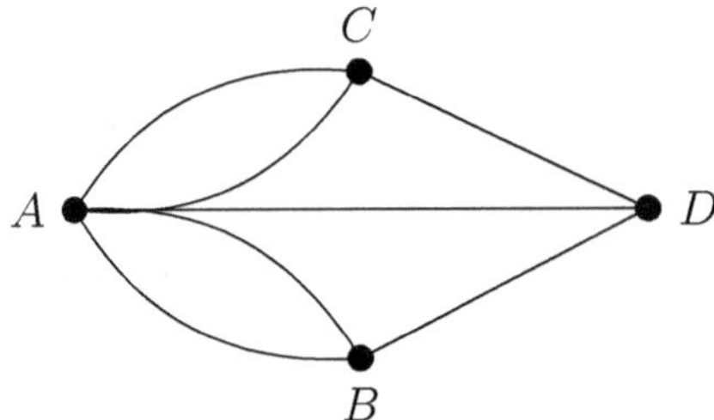

Now looking at the graph we find that all vertices are not of even degree. Hence this graph does not contain an Euler circuit. Hence it is not possible to travel across all bridges exactly once and return to the starting point.

3.1.1   Fleury's algorithm

If the graph contains an Euler circuit (degree of each vertex is even), then we can use Fleury's algorithm to find the Euler circuit.

In this algorithm, we start with any vertex of G and go on tracing the edges of G one by one. While tracing the edges we use the following rules.

   i Erase the edge which is traversed.

   ii Erase the isolated vertex if any.

   iii Traverse the bridge only when there is no alternative.

Finally we reach to a single vertex that is also erased. The edges traced in order give the Euler circuit in G.

Example 3.1. *Find the Euler circuit in the following graph using Fleury's algorithm.*

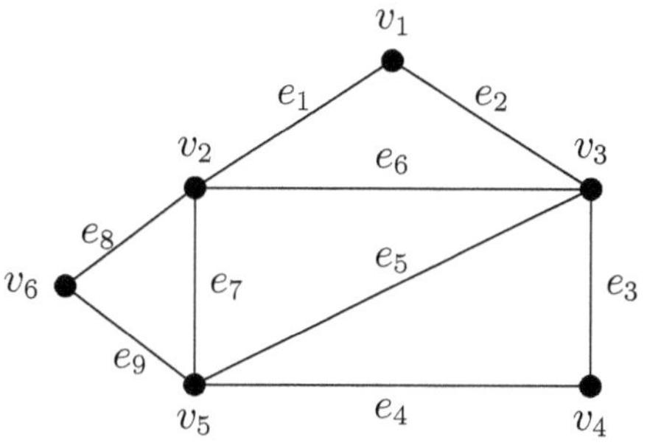

Figure 3.3

Solution: Since the degree of each vertex is even, this graph contains an Euler circuit.

Step	Remaining Graph	Edges traced
Start at v_1 trace edge e_1		
Trace e_8		
Trace e_9		

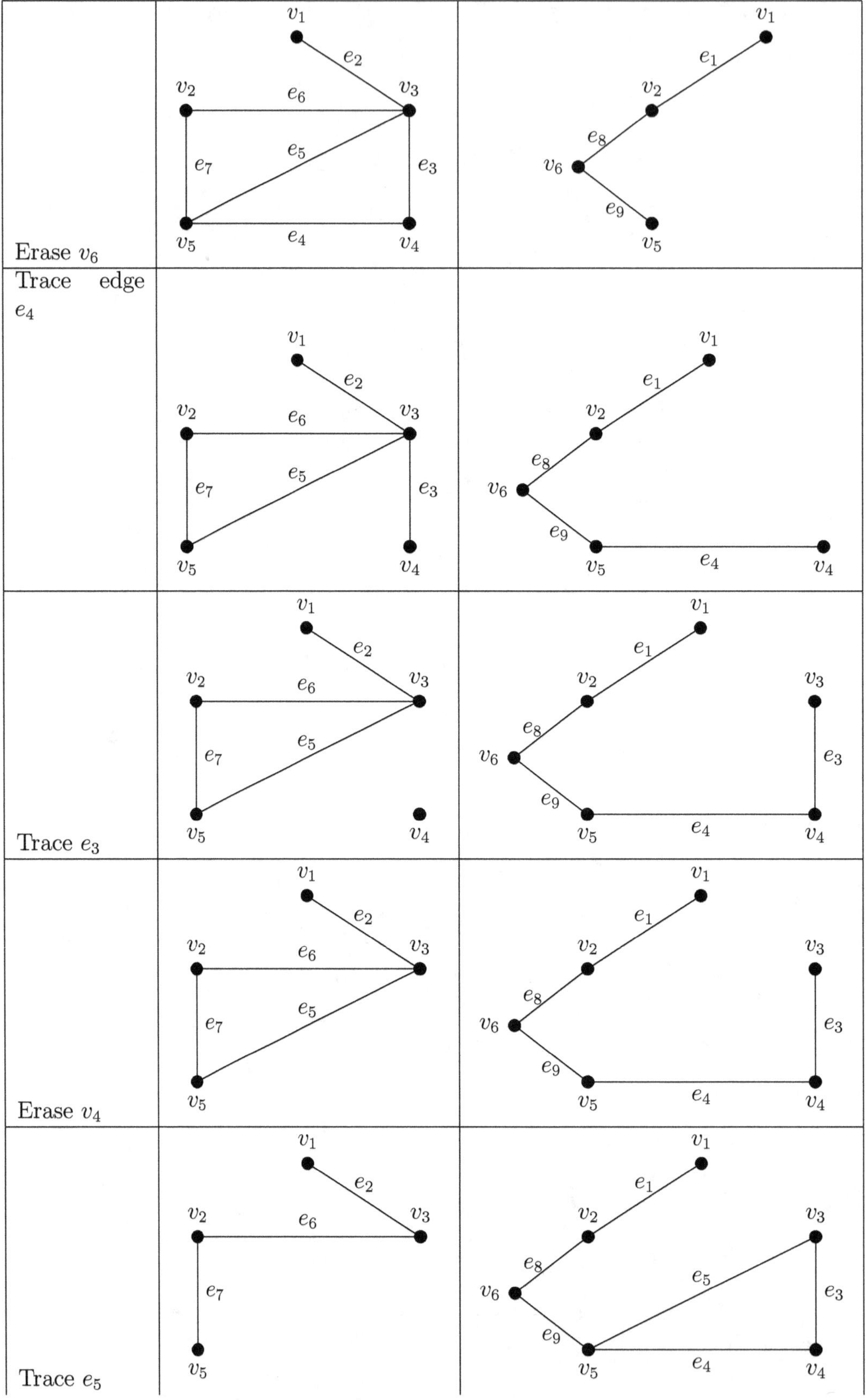
Erase v_6
Trace edge e_4
Trace e_3
Erase v_4
Trace e_5

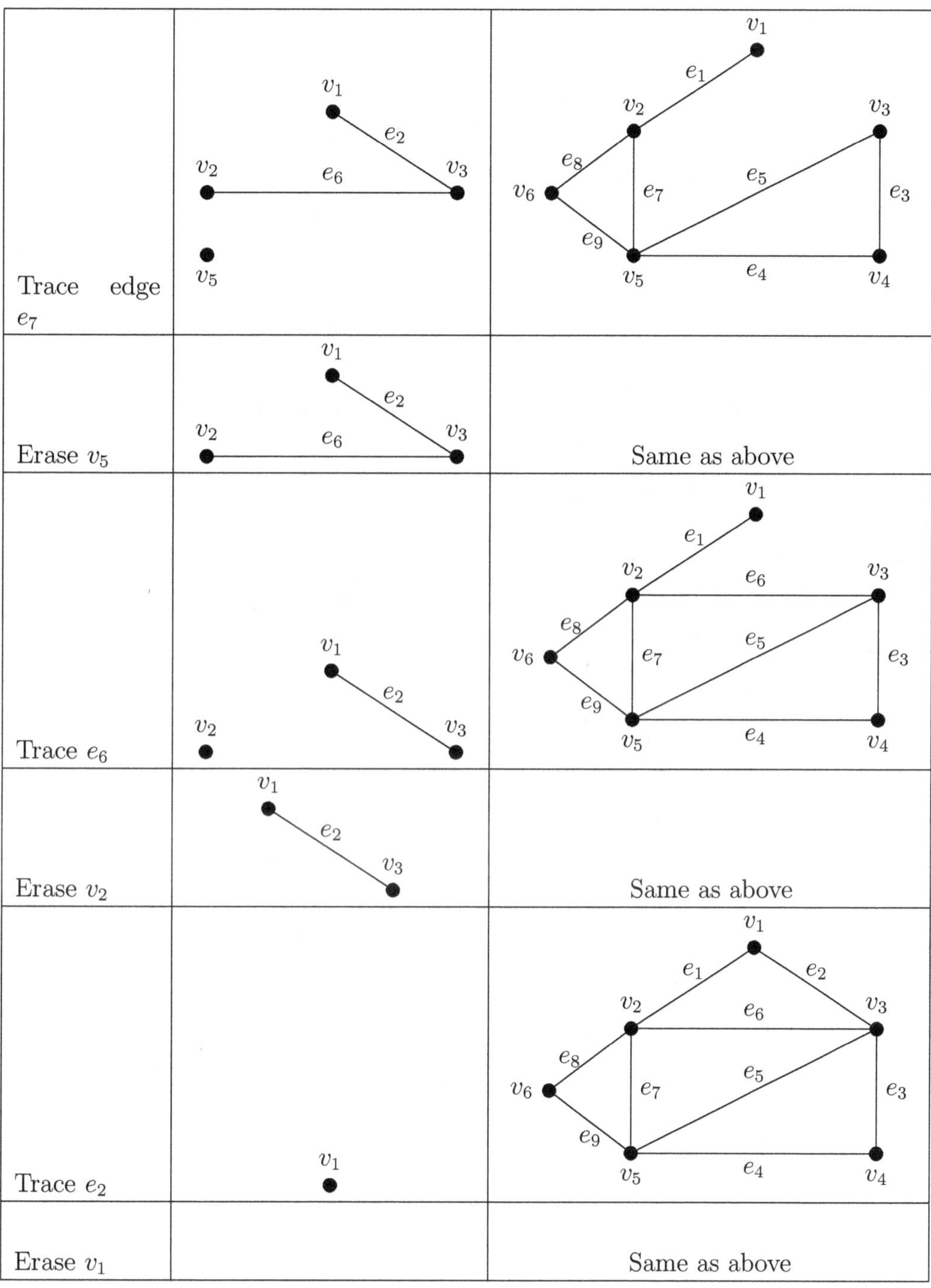

Trace edge e_7		
Erase v_5		Same as above
Trace e_6		
Erase v_2		Same as above
Trace e_2		
Erase v_1		Same as above

Euler circuit is

$$v_1\ e_1\ v_2\ e_8\ v_6\ e_9\ v_5\ e_4\ v_4\ e_3\ v_3\ e_5\ v_5\ e_7\ v_2\ e_6\ v_3\ e_2\ v_1$$

Example 3.2. *Using Fleury's algorithm, find the Euler circuit in the following graph.*

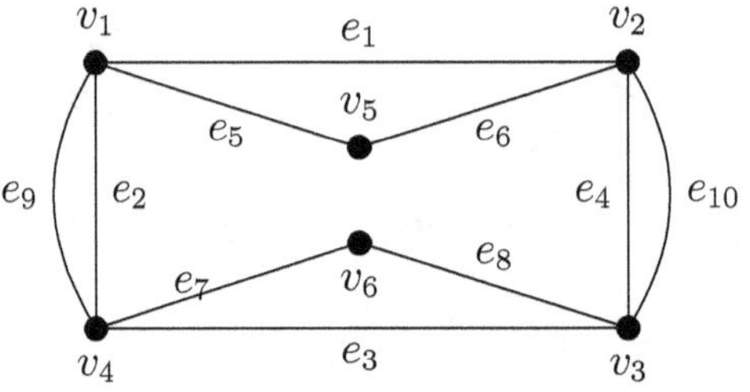

Figure 3.4

Solution: Degree of each vertex is even. Hence we can find Euler circuit in this graph.

Step	Remaining Graph	Edges traced
Start at v_1 trace edge e_1	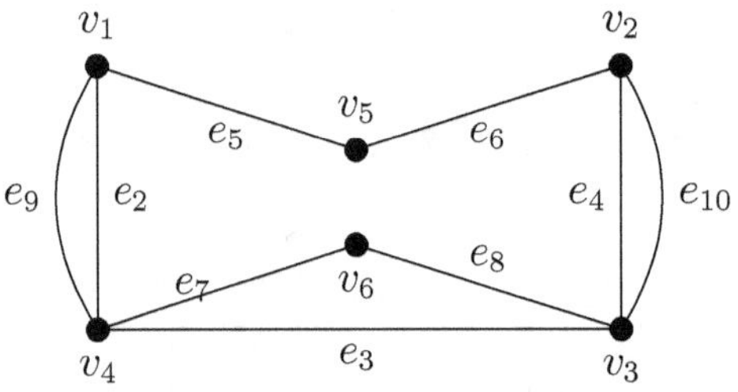	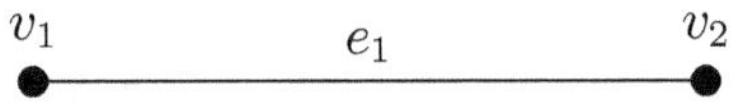
Trace e_4	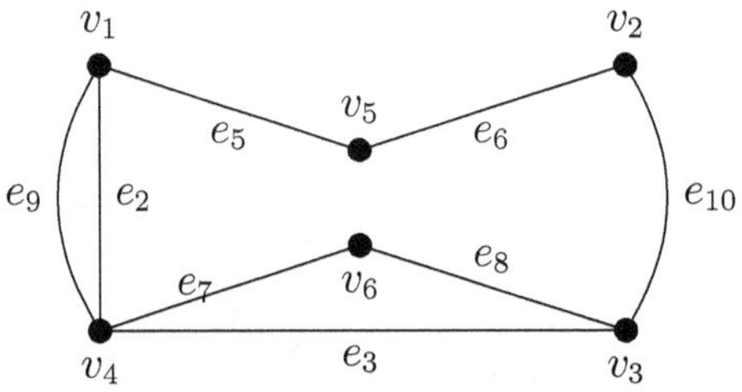	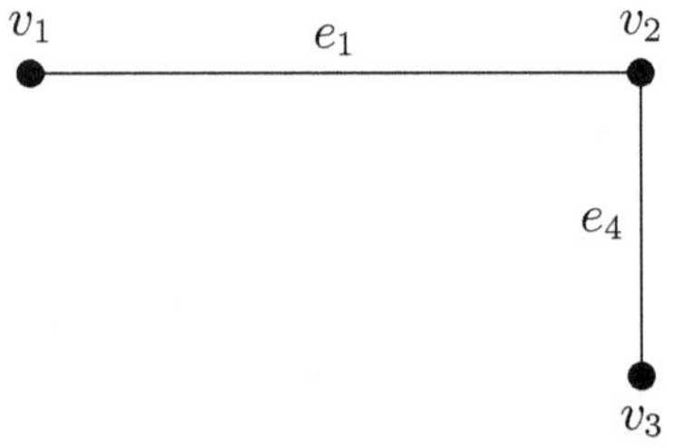
Trace e_3	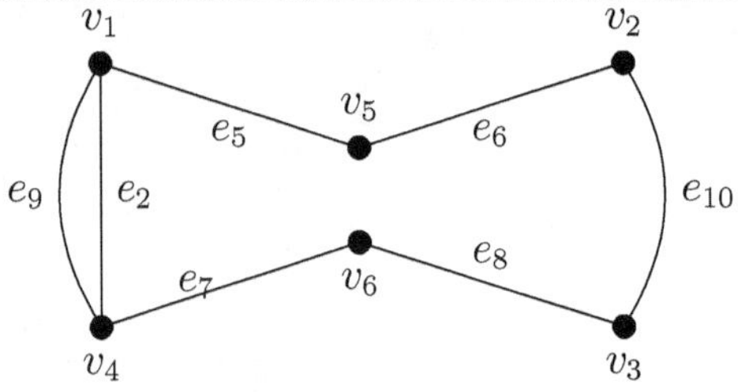	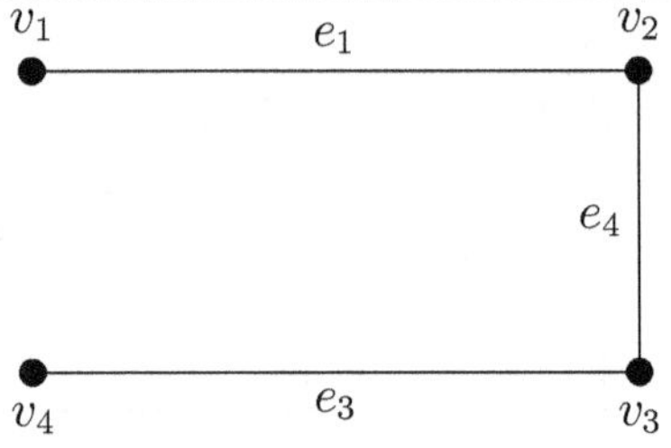

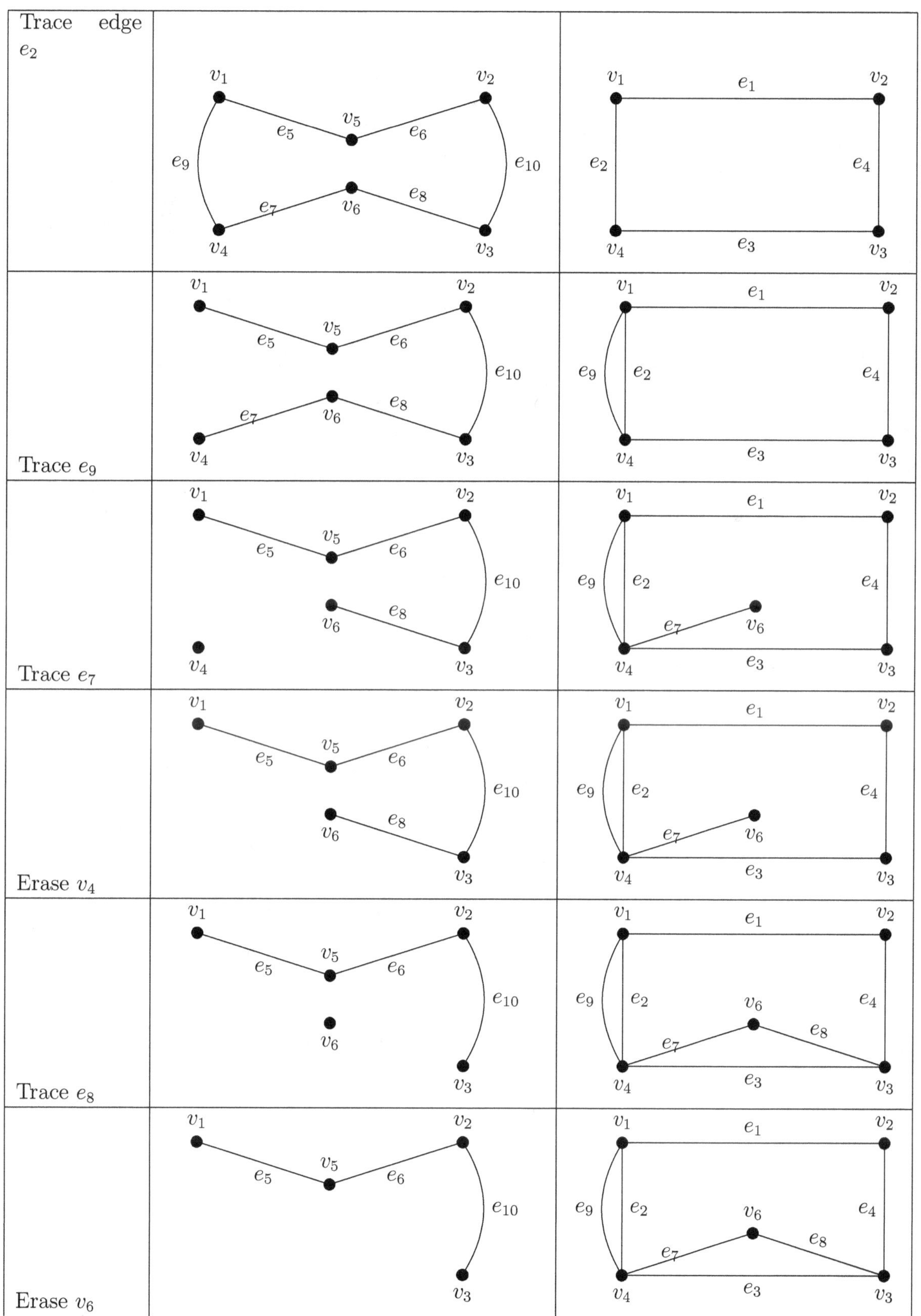

Trace edge e_2
Trace e_9
Trace e_7
Erase v_4
Trace e_8
Erase v_6

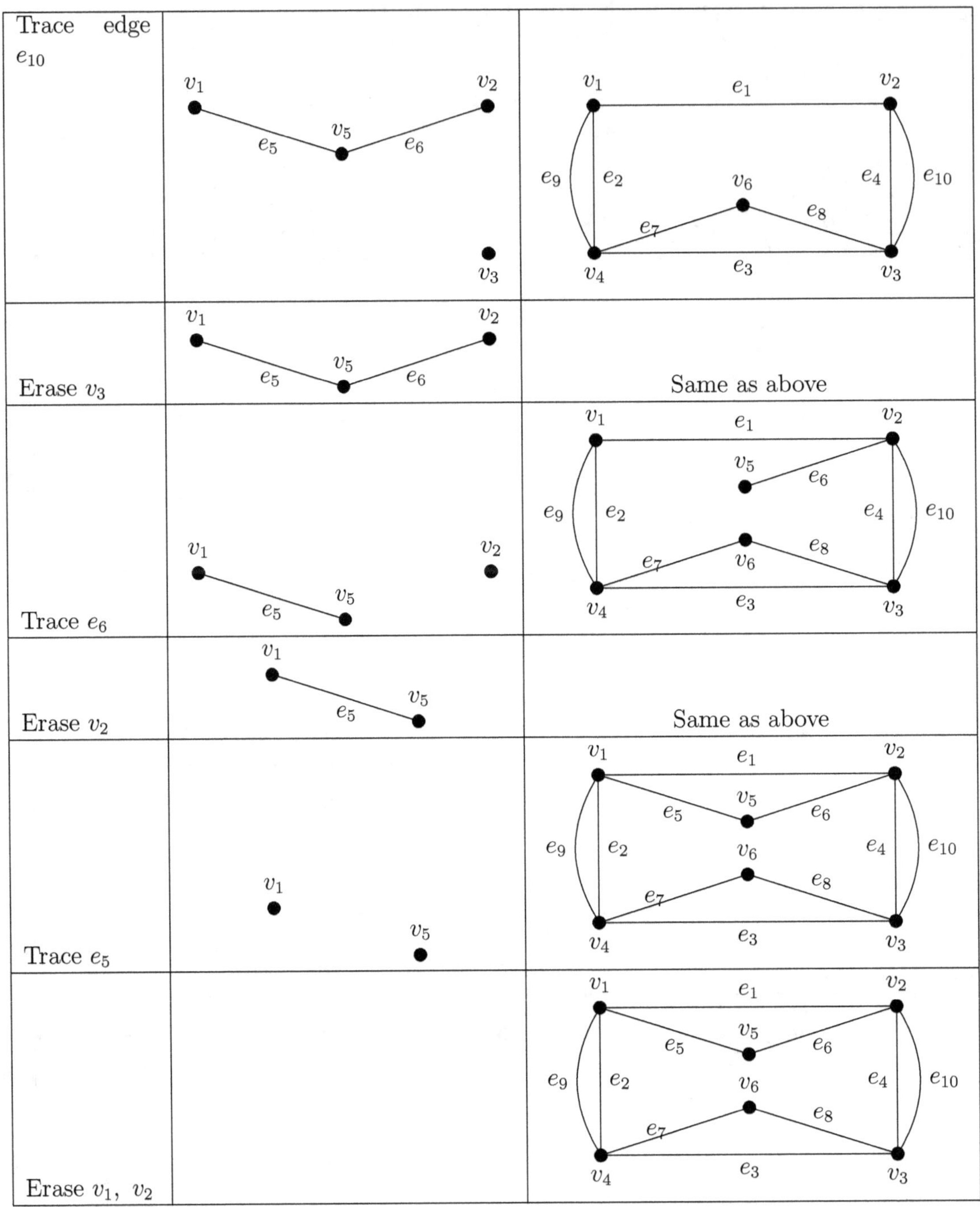

Euler circuit is

$$v_1\ e_1\ v_2\ e_4\ v_3\ e_3\ v_4\ e_2\ v_1\ e_9\ v_4\ e_7\ v_6\ e_8\ v_3\ e_{10}\ v_2\ e_6\ v_5\ e_5\ v_1$$

3.2   Hamilton Path and Circuit

In the previous section we discussed the question of traversing each edge of the graph exactly once. Now we discuss whether can we do the same for the simple path and circuit that contain every vertex of the graph.

Definition 3.4. *Hamilton Path:* *A simple path in a graph G that passes through every vertex exactly once is called a hamilton Path.*

Definition 3.5. *Hamilton Circuit:* *A simple circuit in a graph G that passes through every vertex exactly once is called a hamilton circuit.*

Definition 3.6. *A graph that admits a Hamilton circuit is called Hamiltonian graph.*

Illustrative Example: Consider the graphs given below.

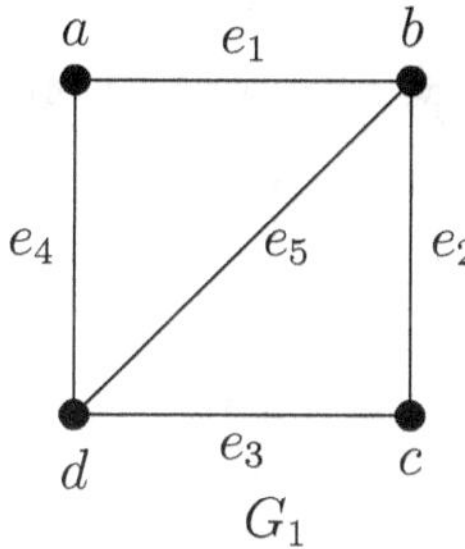
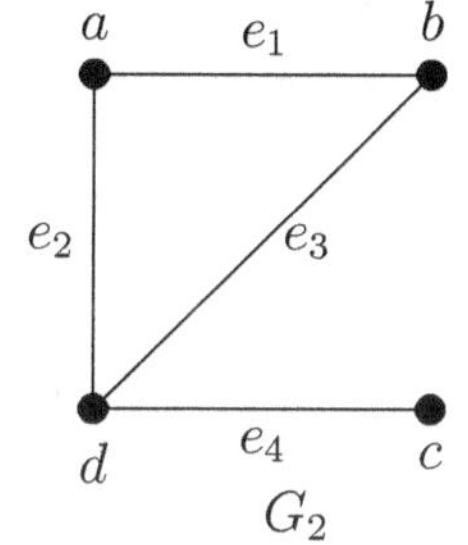
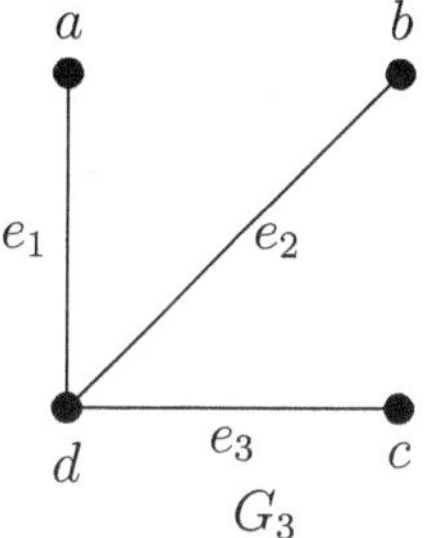

In graph G_1, the Hamilton circuit is $a\ e_1 b\ e_2 c\ e_3 d\ e_4 a$

Graph G_2 has the Hamilton path $b\ e_1 a\ e_2 d\ e_4 c$

Note that graph G_2 does not have Hamilton circuit.   Graph G_3 does not have Hamilton path or Hamilton circuit.

The concept of Hamilton circuit has origin in a game called Icosian puzzle invented in 1857 by the Irish Mathematician Sir William Rowan Hamilton. He made a wooden dodecahedron with a peg at each vertex and a string. Each of the 20 vertices of the dodecahedron was marked with the name of a city. The puzzle was to start at any of the city and travel along the edges of the dodecahedron visiting each of other 19 cities exactly once and return to the first city. The problem reproduces to finding a Hamilton circuit if we represent it graphically. The graph of the dodecahedron with its solution is shown below.

In the following graph the dotted lines give the Hamilton circuit.

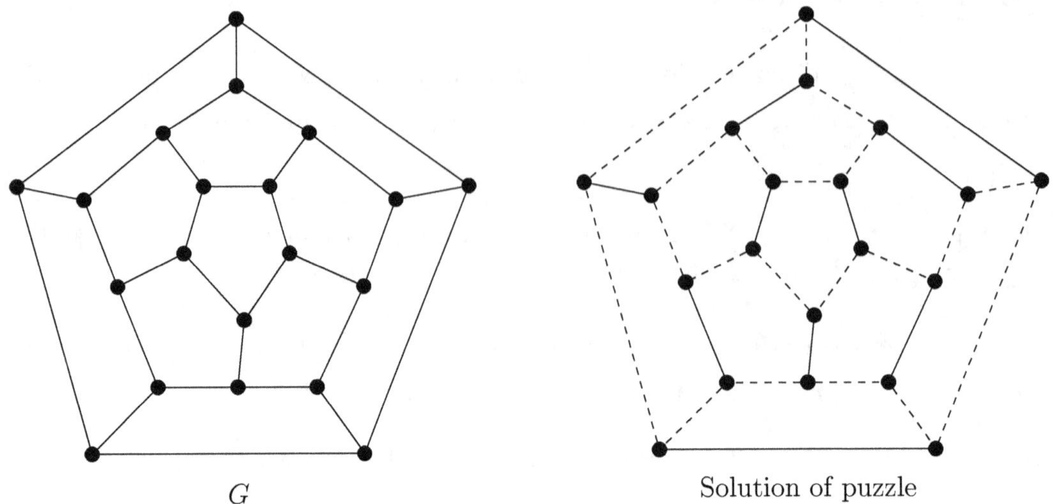

G

Solution of puzzle

Example 3.3. *Show that K_n has a Hamilton circuit whenever $n \geq 3$*

Solution: In K_n, there are edges between any two vertices. So if we start with ay vertex and then visit vertices in any order exactly once then we can return to original vertex. So this is Hamilton circuit in K_n. Hence K_n is Hamiltonion.

Example 3.4. *Show that the following graphs do not have Hamilton circuit.*

i)   ii)

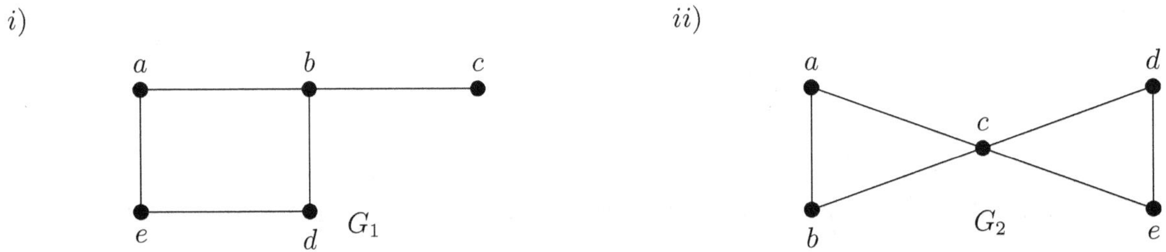

Solution: i) In graph G_1, the vertex c has degree 1. So there does not exists a Hamilton circuit in G_1.

ii) In graph G_2, the vertices a,b,d and e have degree 2. Hence any Hamilton circuit will contain the edges incident on these vertices. So any Hamilton circuit will include all four edges incident on vertex c, which is impossible. Hence G_2 does not have Hamilton circuit.

Example 3.5. *Prove that a bipartite graph with odd umber of vertices can not have a Hamilton circuit.*

Solution: Let $G(V, E)$ be a bipartite graph with $V = V_1 \cup V_2$. Suppose G has a Hamilton circuit. Then it must be of the form $a_1 \, b_1 \, a_2 \, b_2 \cdots a_k \, b_k \, a_1$ where $a_i \in V_1$ and $b_i \in V_2$ for $i = 1, 2, \cdots, k$. Now Hamilton circuit visits every vertex exactly once, except for the first vertex. Therefore the number of vertices is equal to $2k$ which is even. Hence a bipartite graph with odd number of vertices can not

Example 3.6. *For which value of n do the graph K_n has an Euler circuit.*

Solution: Degree each vertex in K_n is $n-1$ For existence of Euler circuit the degree of each vertex must be even, hence $n-1$ must be even i.e. n must be odd.

Example 3.7. *Give an example of a connected graph that has*

   i Neither an Euler circuit nor a Hamilton circuit.

   ii An Euler circuit but no Hamilton circuit.

   iii A Hamilton circuit but not an Euler circuit.

   iv Both Hamilton and Euler circuit.

Solution:

i)

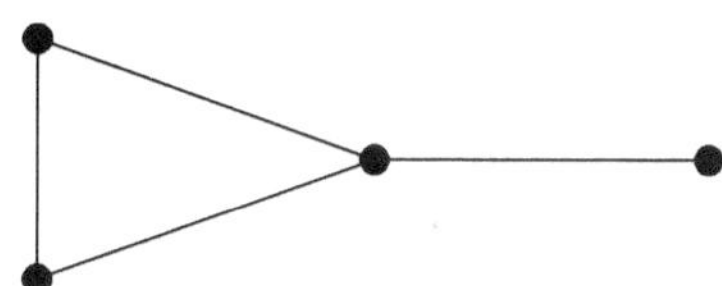

Degree of one vertex is 1. Hence it does not contain a Hamilton circuit. Also all vertices are not of even degree. Hence it does not contain Euler circuit.

ii)

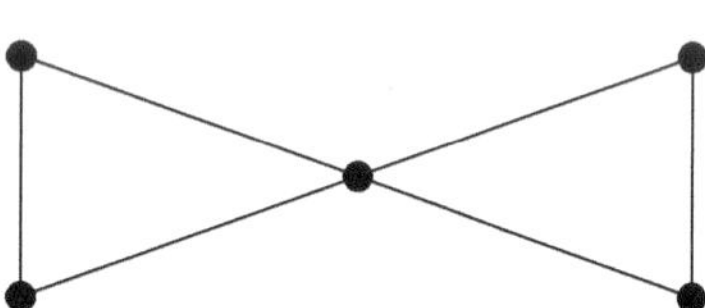

Since degree of each vertex even this is Eulerian graph but this does not contain Hamilton circuit as we have seen before.

iii)

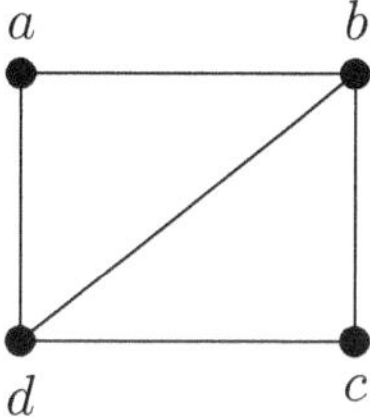

This contains a Hamilton circuit a-b-c-d-a but this has no Euler circuit since two vertices are of odd degree.

iv)

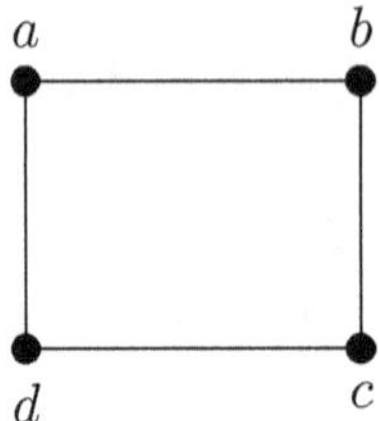

This has both Hamilton and Euler circuit which is nothing but the graph itself.

3.3   Applications

Chinese Postman Problem: The postman is supposed to pick letters from the post office, deliver them along each street on his route and finally return to the post office. While doing so, he would like to cover his route with as little walking as possible. We can represent this situation by means of a graph as follows: The vertices represent junctions of streets. The edges represent the streets and the weights assigned to the edges will be the lengths of the streets between junctions. So the Chinese Postman problem reduces to finding a circuit containing every edge of weighted graph with minimum weight.

If the graph is Eulerian, then the Euler circuit is the solution to the problem. If the graph is not Eulerian we duplicate some edges along the shortest path between the vertices of odd degree, to obtain a new graph which is Eulerian. In this new graph, we can find Euler circuit. Practically this will mean that we have to repeat some of the streets.

Example 3.8. *Solve the following Chinese postman problem.*

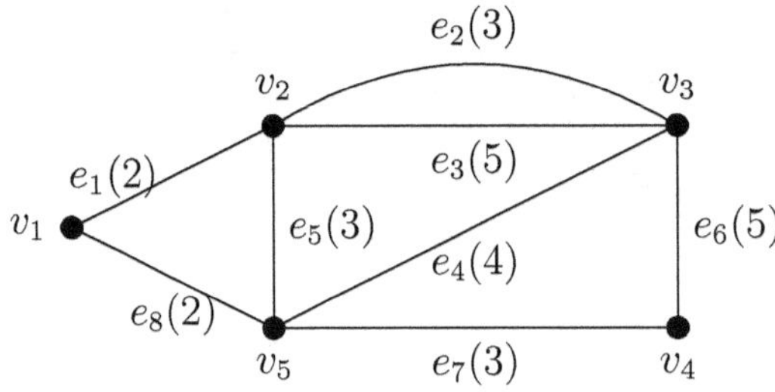

Figure 3.5

Solution: Since the degree of each vertex is even, the graph contains an Euler circuit which is the solution to this problem. Using Fleury's algorithm, we can get the following Euler circuit.

$$v_1 \ e_1 \ v_2 \ e_2 \ v_3 \ e_6 \ v_4 \ e_7 \ v_5 \ e_4 \ v_3 \ e_3 \ v_2 \ e_5 \ v_5 \ e_8 \ v_1$$

Weight of Euler circuit $= 2 + 3 + 5 + 34 + 5 + 3 + 2 = 27$.

Example 3.9. *Solve the following Chinese postman problem.*

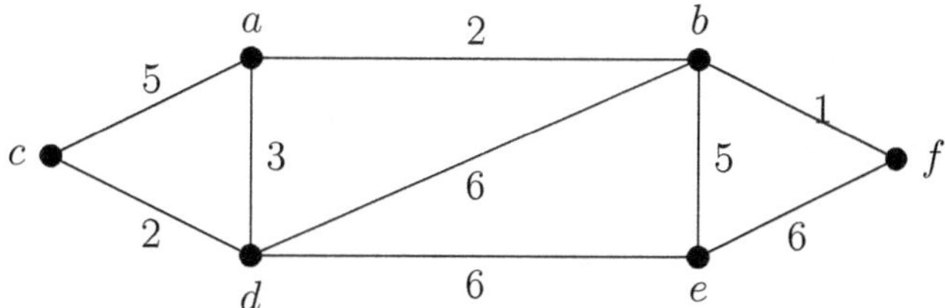

Figure 3.6

Solution: Given graph does not contain an Euler circuit since it contains two vertices viz 'a' and 'e' with odd degree. The shortest route between vertices 'a' and 'e' is via vertex 'b'. So we duplicate the edges along this route. Then the new graph is

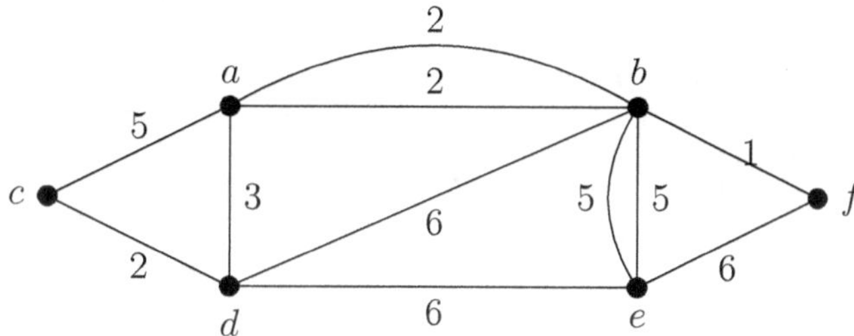

Now we can find Euler circuit using Fleury's algorithm. the Euler circuit is

a-b-c-f-b-c-d-b-a-d-e-a.

Weight of the Euler circuit is

$2 + 5 + 6 + 1 + 5 + 6 + 6 + 2 + 3 + 2 + 5 = 43.$

Travelling Salesman Problem : Suppose a travelling salesman's territory includes several towns with roads connecting certain pairs of towns. Travelling Salesman Problem is to visit each of towns exactly once and return to the starting point so that the salesman has to travel the minimum total distance. If we represent town by vertices and roads by edges then this problem asks for a Hamilton circuit with minimum total weight. There is no efficient available algorithm to find Hamilton circuit of least length. The most straight forward way to solve this is to examine all possible Hamilton circuits and select one of minimum total length.

Example 3.10. *Solve the following Travelling Salesman problem.*

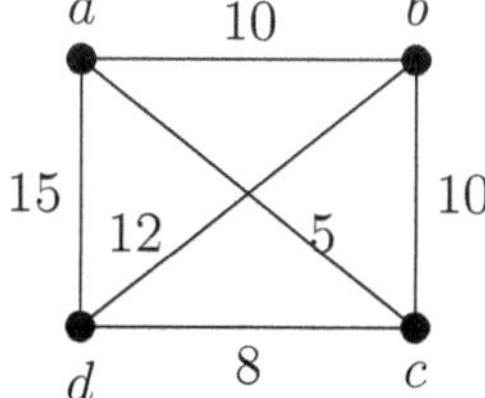

Solution: There are three possible Hamilton circuits as shown below.

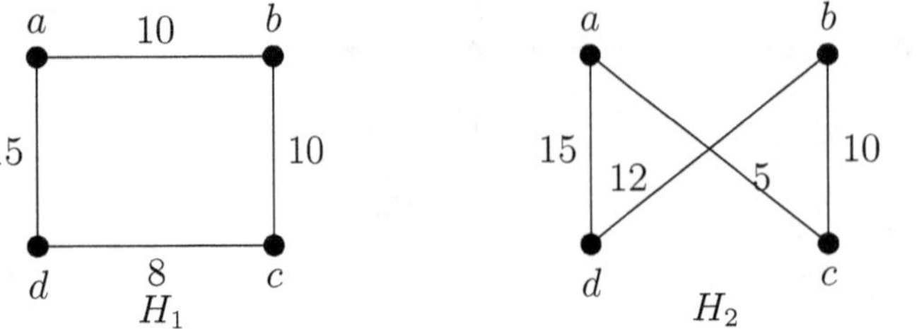
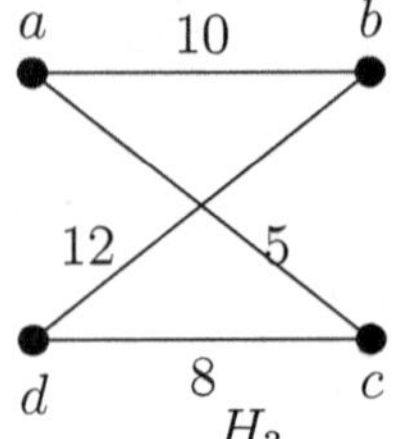

Weight of Hamilton circuit $H_1 = 10 + 10 + 15 + 8 = 43$.

Weight of Hamilton circuit $H_2 = 15 + 12 + 5 + 10 = 42$.

Weight of Hamilton circuit $H_3 = 10 + 8 + 12 + 5 = 35$.

So the solution to the problem is the Hamilton circuit H_3 with minimum weight 35.

Example 3.11. *Solve the following Travelling Salesman problem.*

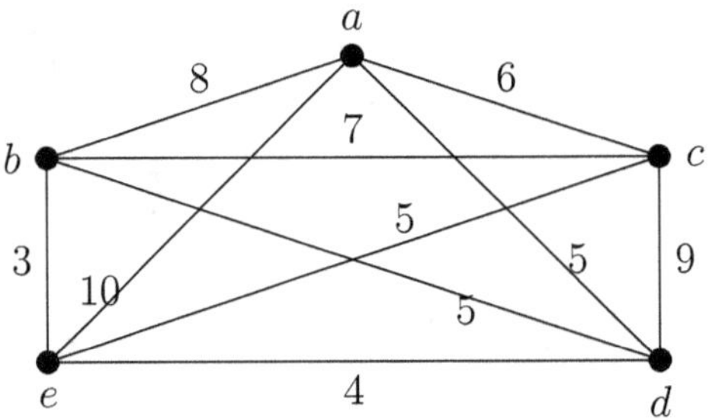

Figure 3.8

Solution: We will start with vertex 'a' and will choose nearest neighbour each time from non selected vertices.

Nearest neighbour of 'a' is 'd' (incident edges have weights $8, 7, 5, 6$)

$$a - d$$

Nearest neighbour of 'd' is 'e' (incident edges have weights $4, 5, 9$)

$$a - d - e$$

Nearest neighbour of 'e' is 'b' (incident edges have weights $3, 10, 5$)

$$a - d - e - b$$

Now last vertex is 'c' and then we return to 'a'.

Weight of the Hamilton circuit is $5 + 4 + 3 + 7 + 6 = 25$.

Exerecise

1. Show that the following graphs have Euler circuit. Using Fleury's algorithm, find the Euler circuit.

$i)$

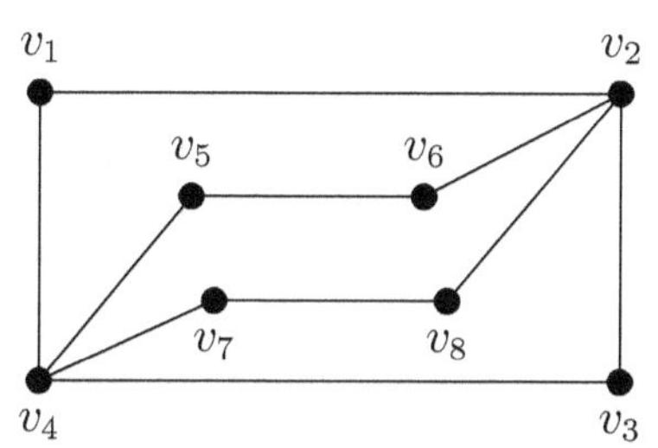

$ii)$

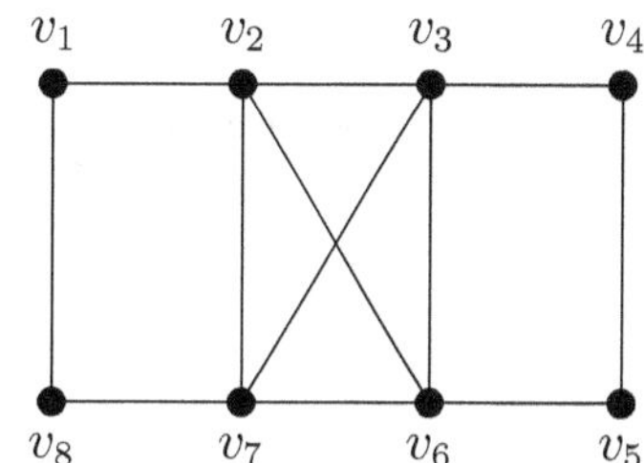

2. Find Euler paths in the following graphs.

$i)$ 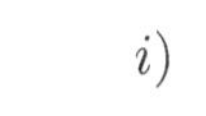

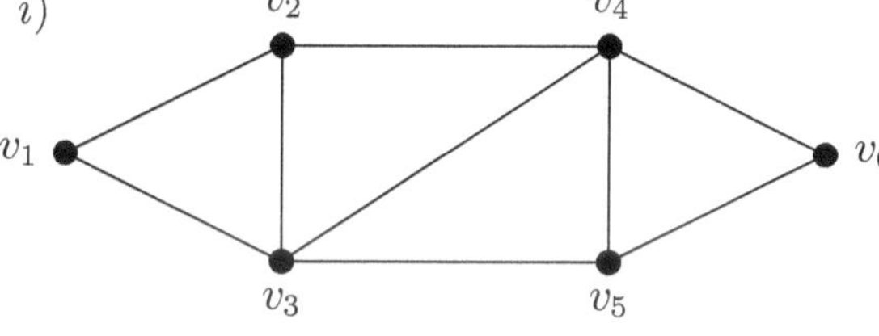

$ii)$

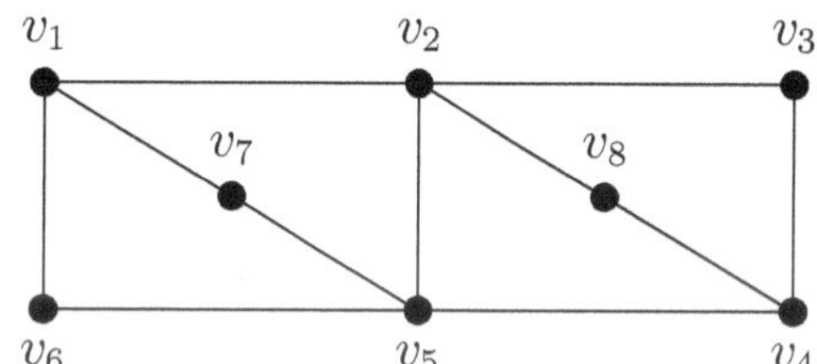

3. Find the Hamilton circuit in the following graphs.

$i)$

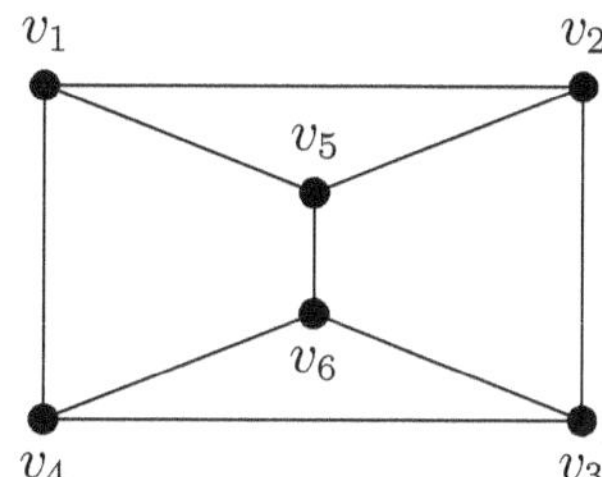

$ii)$

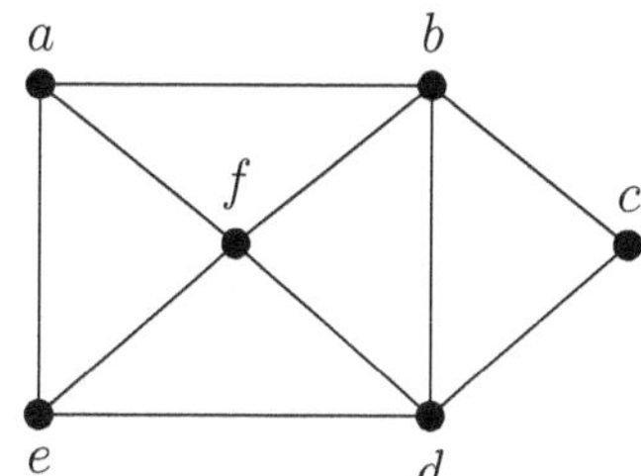

4. Check whether the following graph is a) Eulerian b) Hamiltonian. Justify.

$i)$

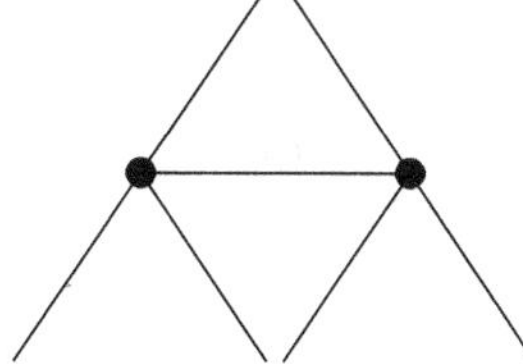

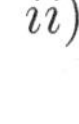

$ii)$

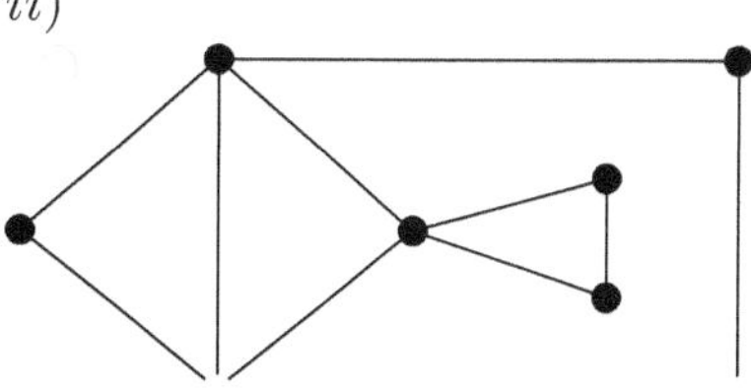

5. Solve the following Chinese Postman problem

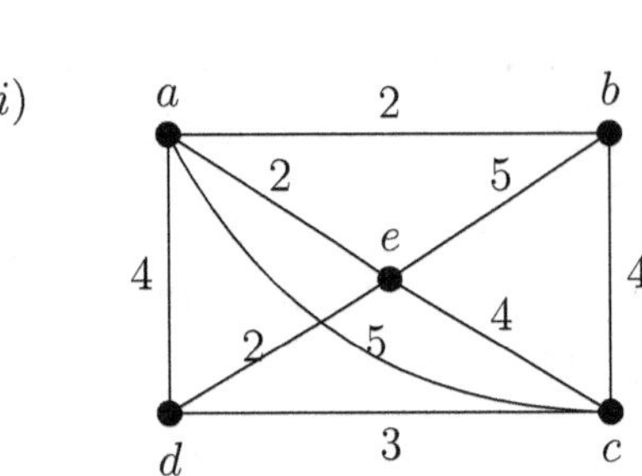

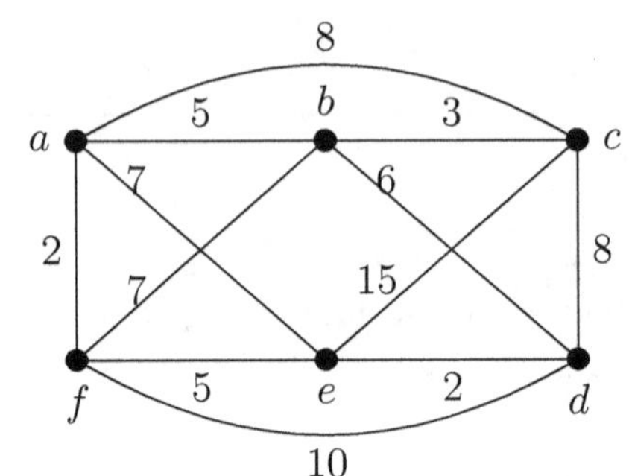

6. Solve the following Travelling Salesman problem.

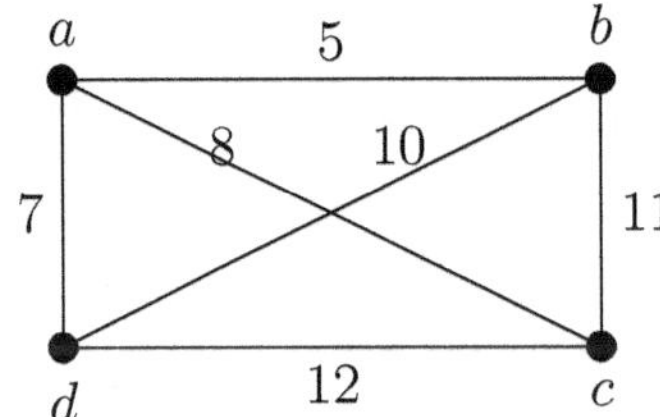

Solutions

1.   i $v_1 - v_2 - v_6 - v_5 - v_4 - v_7 - v_8 - v_2 - v_3 - v_4 - v_1$

     ii $v_1 - v_2 - v_3 - v_4 - v_5 - v_6 - v_3 - v_7 - v_6 - v_2 - v_7 - v_8 - v_1$

2.   i $v_2 - v_1 - v_3 - v_5 - v_6 - v_4 - v_2 - v_3 - v_4 - v_5$

     ii $v_1 - v_2 - v_5 - v_6 - v_1 - v_7 - v_5 - v_4 - v_3 - v_2 - v_8 - v_4$

3.   i $v_1 - v_5 - v_6 - v_4 - v_3 - v_2 - v_1$

     ii $a - f - e - d - c - b - a$

4.   i Eulerian as well as Hamiltonian

    ii Eulerian but not Hamiltonian

5.   i) Add edges along shortest path between vertices b and d to make it Eulerian. The new graph is

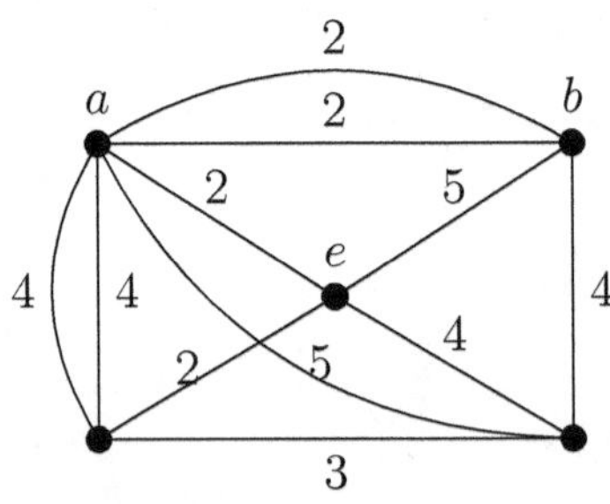

Euler circuit is $a - b - c - e - b - a - e - d - c - a - d - a$ whose weight is 37.

ii) The graph is Eulerian. Euler circuit is

$$a - b - c - d - f - e - d - b - f - a - e - c - a$$

Weight $= 10 + 5 + 3 + 2 + 8 + 5 + 3 + 8 + 7 + 7 + 6 + 15 = 79$.

6. Three possible circuits.

$a - b - c - d - a$ weight $= 35$

$a - d - b - c - a$ weight$= 36$

$a - b - d - c - a$ weight $= 35$

So solution is any circuit with weight 35.

Chapter 4

Trees

Introduction: Tree is a special type of connected graph discovered by kirchoff in 1847, in order to solve system of simultaneous linear equations which give the current in each branch and around each circuit of an electric network. In 1857, Cayley used trees to count certain types of chemical compounds. As a graph, trees are widely used in data storage, searching and communication. Trees play an important role in a variety of algorithms, games and puzzles.

4.1   Definition and Properties of a tree

Definition 4.1 (Tree:). *A connected graph that contains no simple circuit is called a tree.*

Note that a tree cannot contain multiple edges or loops. Hence a tree is a simple graph. The following figure 4.1 shows trees on $1, 2, 3, 4$ and 5 vertices respectiveley.

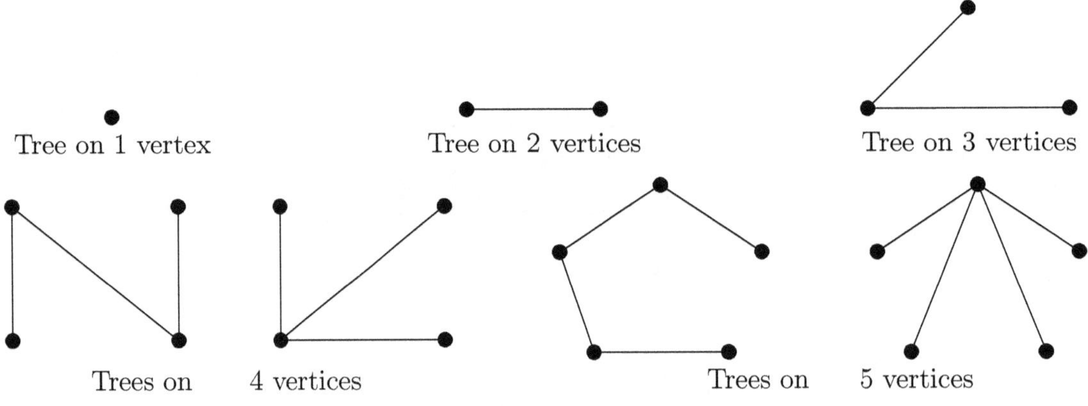

Figure 4.1

Theorem 4.1. *A graph T is a tree if and only if there is exactly one simple path between any two of its vertices.*

Proof. Suppose that T is a tree. So T is a connected graph without any circuit. Let u and v be two vertices of T. Then by connectedness of T, there exists a simple path between u and v. To prove uniqueness, if possible suppose that there are two different paths between u and v. Now the path formed by combining the first path from u to v followed by the path from v to u by reversing the order of second path from u to v would form a circuit. This is contradiction, since T is circuit free. Hence there exists exactly one simple path between any two vertices of T.

Conversely suppose that there is exactly one simple path between any two vertices of a graph T. The existence of a path between any two vertices proves that T is connected. Now we prove that, T cannot have circuit. If possible, suppose T has a circuit and let u and v be vertices belonging to this circuit. Then there would be two different paths between u and v; which is contradiction to the assumption that there is exactly one path between any two vertices. Hence T is circuit free. Therefore T is a tree according to the definition. $\square$

Theorem 4.2. *A tree with n vertices has $n - 1$ edges.*

Proof. We use induction on number of vertices to prove the theorem. If $n = 1$, then T has only one vertex. Since T does not contain self loop, T will have no edges. That is, it has $n - 1 = 1 - 1 = 0$ edges. So the result is true for n=1.

Now suppose that the result holds for $n = k$. That is a tree with k vertices has $k - 1$ edges. Now let T be a tree with $k + 1$ vertices. Let u be a vertex of degree 1 in T and let 'e' be the edge having one end vertex u. After deleting edge 'e', we get exactly two components namely $T - u$ and u. Note that $T - u$ is a circuit free and connected graph and hence a tree. Since $T - u$ contains k vertices, by assumption, it has $k - 1$ edges. Now T has exactly one extra edge than $T - u$. Hence T has $(k - 1) + 1 = k$ edges as required.

Thus, by principle of mathematical induction the result holds for all positive integers k. $\square$

Theorem 4.3. *A graph G with n vertices, $n - 1$ edges and no circuits is a tree.*

Proof. Since G is a circuit free, it remains to prove that G is connected. If possible, suppose G is disconnected, then G will contain two or more connected components without loss of generality, suppose G has two connected components say G_1 and G_2 with vertices n_1 and n_2 respectively. Now G_1 is connected and circuit free and hence is a tree. So G_1 contains $n_1 - 1$ edges. By similar argument G_2 contains $n_2 - 1$ edges. So total edges in G will be $(n_1 - 1) + (n_2 - 1) = (n_1 + n_2) - 2 = n - 2$ which is less than $n - 1$. This is a contradiction. Hence G must be connected. Now G being a circuit free and connected graph, it is a tree. $\square$

Theorem 4.4. *A tree with two or more vertices has at least two pendant vertices.*

Proof. Let T be a tree with n vertices, where $n \geq 2$. Since the number of edges in T is $n - 1$, the total degree of T is $2(n - 1) = (2n - 2)$. Since T is connected, degree of each vertex is at least one. If we assign one degree to each vertex, there remains $(2n - 2) - n = n - 2$ degrees. While distributing vertices have only degree 1 assigned in the first round of distribution and hence they are pendant vertices of T. $\qquad\square$

4.2 Centre of a tree

Definition 4.2. *Eccentricity of a vertex: Let T be a tree and let v be any fixed vertex in T. The distance of the farthest vertex from v is called as eccentricity of v in T.*

Eccentricity of a vertex v is denoted by $E(v)$.

$$E(v) \; = \; \max_{v_i \in T} \{d(v, v_i)\}$$

Definition 4.3. *Centre of a tree: A vertex with minimum eccentricity is called as centre of a tree.*

Definition 4.4. *Radius and Diameter of a tree: The eccentricity of the centre of a tree is called as radius of a tree. Alternatively, the minimum value of the eccentricity is called as radius of a tree. Also the maximum value of the eccentricity is called as diameter of a tree.*

Example 4.1. *Find centre, radius and diameter of the following tree.*

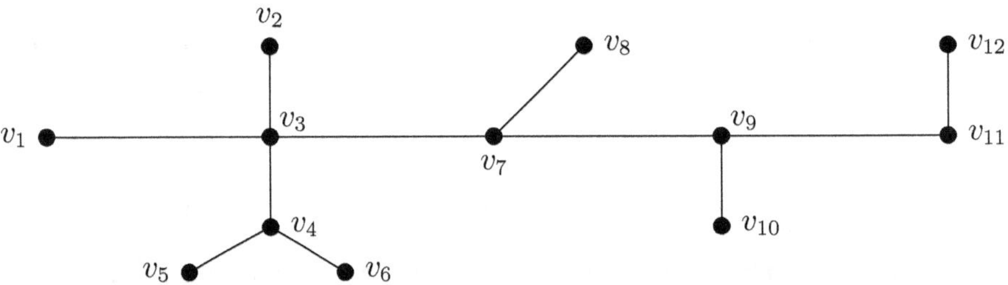

Figure 4.2

Solution: Note that, distance of vertex v_1 from vertex v_{12} is 5, and distance of v_1 from any other vertex is less than 5. Hence eccentricity of v_1 is 5. So $E(v_1) = 5$

Similarly we find the eccentricities of other vertices.

$E(v_2) = 5$, $E(v_7) = 3$, $E(v_3) = 4$, $E(v_8) = 4$, $E(v_4) = 5$, $E(v_9) = 4$, $E(v_5) = 6$, $E(v_{10}) = 5$, $E(v_6) = 6$, $E(v_{11}) = 5$, $E(v_{12}) = 6$.

Vertex v_7 has minimum eccentricity. Hence centre of the tree is v_7.

Radius of the tree is the minimum eccentricity value i.e. 3.

Diameter of the tree is the maximum eccentricity value i.e. 6.

Example 4.2. *Find centre, radius and diameter of the following tree.*

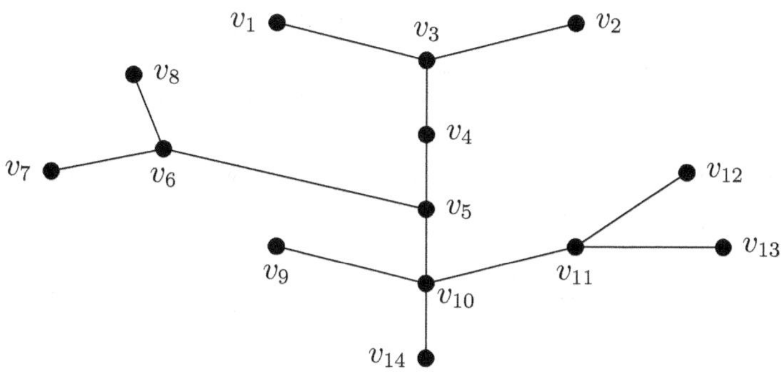

Figure 4.3

Solution:

$E(v_1) = 6, \quad E(v_2) = 6, \quad E(v_3) = 5, \quad E(v_4) = 4$

$E(v_5) = 3, \quad E(v_6) = 4, \quad E(v_7) = 5, \quad E(v_8) = 5,$

$E(v_9) = 5, \quad E(v_{10}) = 4, \quad E(v_{11}) = 5, \quad E(v_{12}) = 6, \quad E(v_{13}) = 6, \quad E(v_{14}) = 5.$

Hence centre is v_5.

Radius of the tree is 3.

Diameter of the tree is 6.

Example 4.3. *Find Centre and radius of the following tree.*

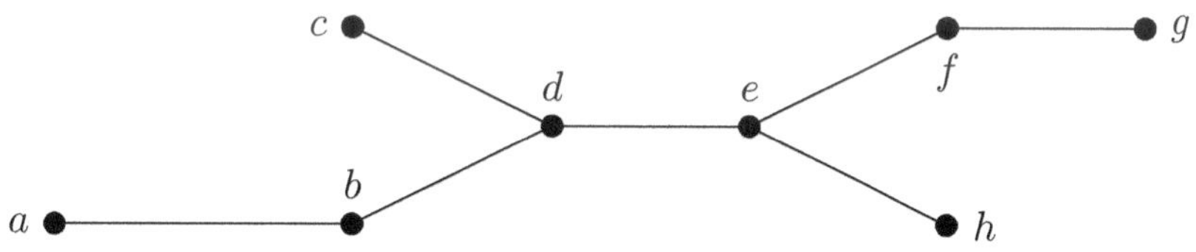

Figure 4.4

Solution:

$E(a) = 5 \qquad\qquad E(e) = 3$

$E(b) = 4 \qquad\qquad E(f) = 4$

$E(c) = 4 \qquad\qquad E(g) = 5$

$E(d) = 3 \qquad\qquad E(h) = 4$

Centres of tree are vertices d and e.

Radius of the tree is 3 and diameter of the tree is 5.

Theorem 4.5. *Every tree has a centre which consists of either a single vertex or two adjacent vertices.*

Proof. If the tree is on one or two vertices then the result holds. So suppose that the tree has more

by removing all pendant vertices from T. We now prove that the centre of the T and T_1 are same. For this note that, after deleting pendant vertices, eccentricity of each vertex reduces by 1. Therefore the vertex with minimum eccentricity remain same in T and T_1. Hence centers of T and T_1 are same. From T_1, we again remove all pendant vertices to get a tree T_2. Now by similar argument as before centers of T, T_1 and T_2 are same. Continuing this procedure, we obtain either a single vertex k_1 or two adjacent vertex k_2 which are nothing but centers of T. $\qquad\square$

In the next example, we will use the procedure mentioned in the above theorem to find centre of a tree.

Example 4.4. *Find centre of the following tree.*

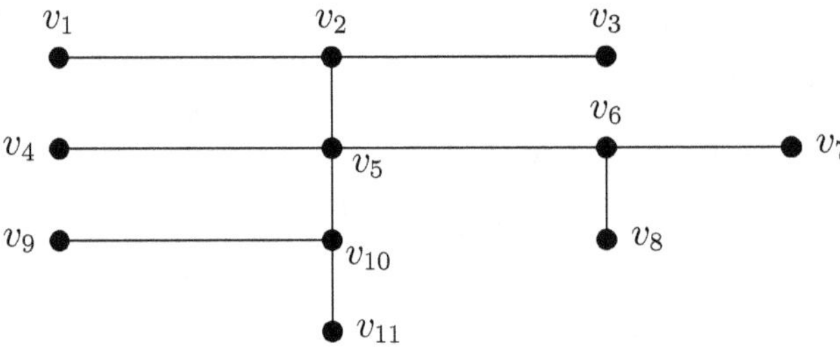

Figure 4.5

Solution: Step1: Removing all pendant vertices from given tree we get following tree T_1.

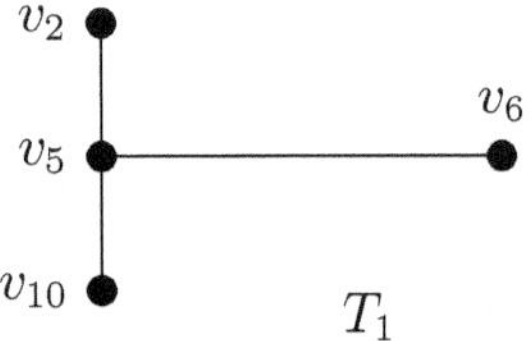

Again removing pendant vertices from T_1, we get T_2.

Therefore centre of the tree is $'v_5'$.

4.3 Spanning trees:

Definition 4.5. *Let G be a simple graph. A spanning tree of G is a subgraph of G, that is a tree containing every vertex of G.*

Example 4.5. *Find the spanning tree of the graph G shown below.*

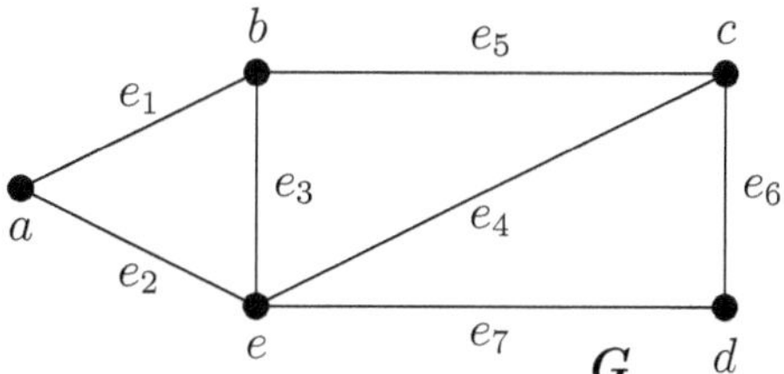

Figure 4.6

Solution: Given graph G is with 5 vertices and 7 edges. Since a tree on 5 vertices will contain only 4 edges, we have to remove 3 edges from G such that the graph remains connected. Obviously, we have to remove edges from circuits. If we remove e_1, e_5 and e_7, we get following spanning tree.

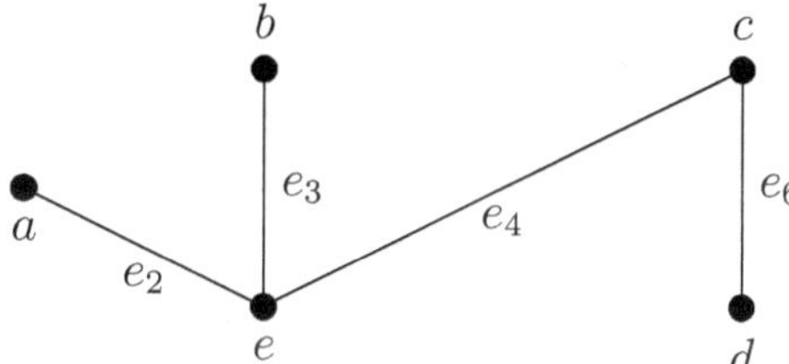

Theorem 4.6. *A simple graph G is connected if and only if it has a spanning tree.*

Proof. First we suppose that G has a spanning tree. Then T contains every vertex of G. There is a path in T between any two of its vertices. But T being a subgraph of G, there is a path in G between any two vertices. Hence G is connected.

Now we suppose that G is connected. If G does not contain any circuit then G will be a tree and will be a spanning tree as required. Suppose G contains a simple circuit. We remove one edge from this circuit. The resulting subgraph contains all vertices of G. Moreover it is still connected, because when two vertices are connected by a path containing the removed edge, they are connected by an alternative path of the circuit not containing this edge. If this subgraph is not a tree, it must contain some circuit. So as above we repeat the procedure of removing edge from a circuit until the graph becomes circuit free. At the end we will get a circuit free connected graph containing all vertices of G which will be a spanning tree of G. $\square$

Definition 4.6. ***Chords and Branches of a tree:*** *Let G be a connected graph and T be its spanning tree. Then the edges of G which are included in the spanning tree T are called branches of T, while the remaining edges of G are called chords of T.*

Example 4.6. *Consider the following graph G and its spanning tree T.*

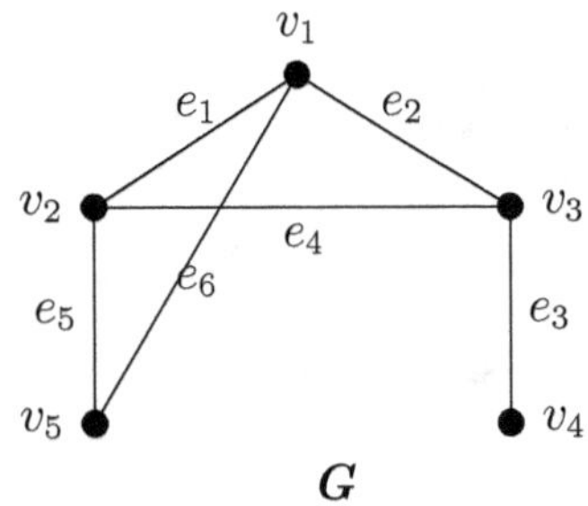

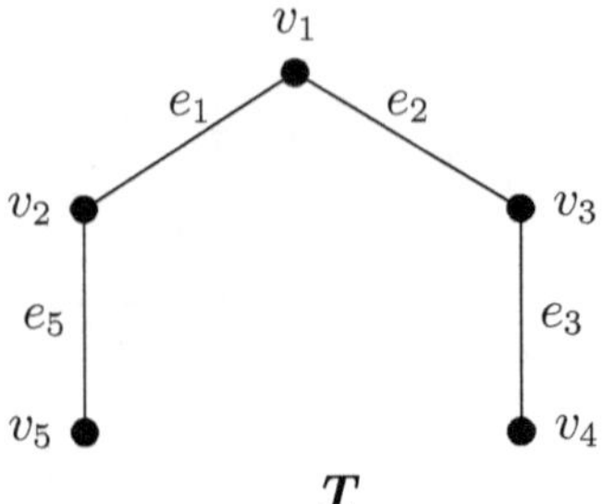

Figure 4.7

Find branches and chords of G with respect to spanning tree T.

Solution: Branches of T: e_1, e_2, e_3, e_5

Chords of T: e_4, e_6.

Remark 4.1. *Note that the terms branch and chord are relative terms. A branch (or chord) with respect to one spanning tree may not be a branch (or chord) with respect to another spanning tree. But the number of branches and chords is fixed for a given graph. if a graph G on n vertices contains e edges, then it contains $n - 1$ branches and $e - (n - 1)$ chords.*

Example 4.7. *Draw any four spanning trees of the following graph.*

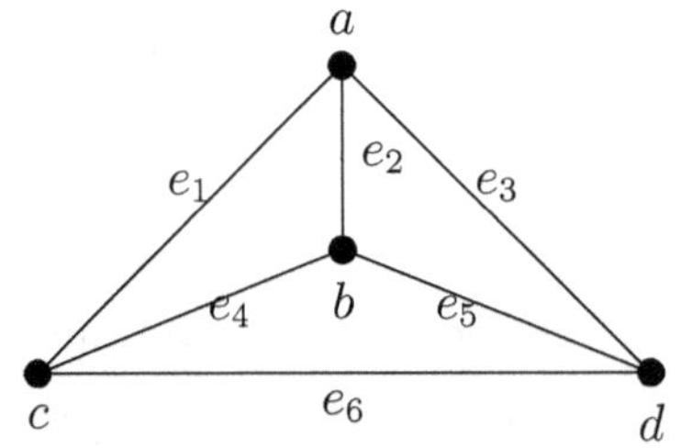

Figure 4.8

Solution:

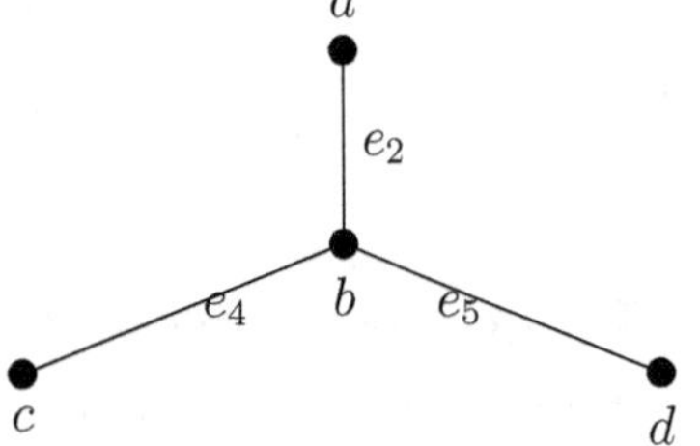

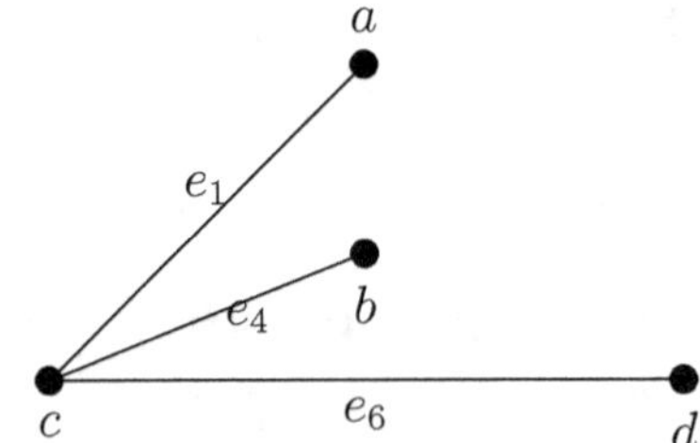

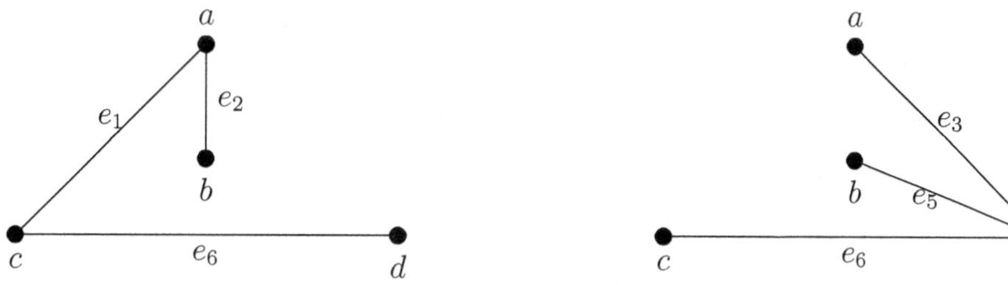

Weighted graphs: Many problems can be modeled using graphs with weights assigned to their edges. For example, if the vertices denote the cities and edges represent the roads connecting them, then the weights assigned to the edges may represent distance between the two cities. In such problem, we require to find a path of least length between any two vertices in a network. This can be solved if we find spanning tree of the given graph having least weight. In this section, we will see two methods for finding such minimal or shortest spanning tree.

Definition 4.7. *Weighted graph: A graph G in which each edge is assigned a number is called a weighted graph.*

The number assigned to the edge 'e' is called as weight of the edge e and is denoted by $w(e)$. The sum of the weights of all edges in a graph G is called as weight of G and is denoted by $w(G)$.

Example 4.8. *For the following weighted graph, draw any two of its spanning trees. Also write their weights.*

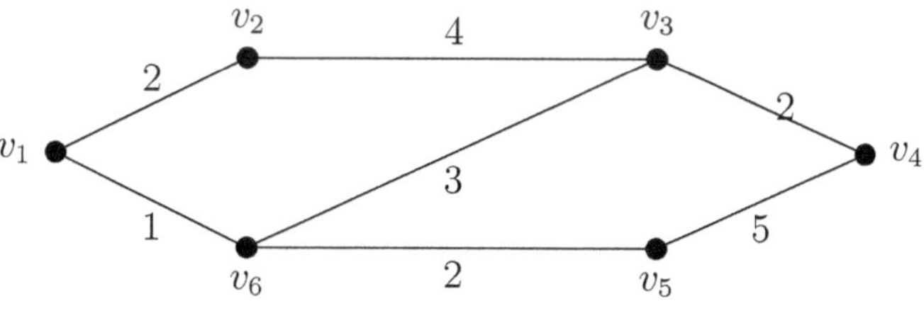

Figure 4.9

Solution: The trees T_1 and T_2 shown below are the spanning trees.

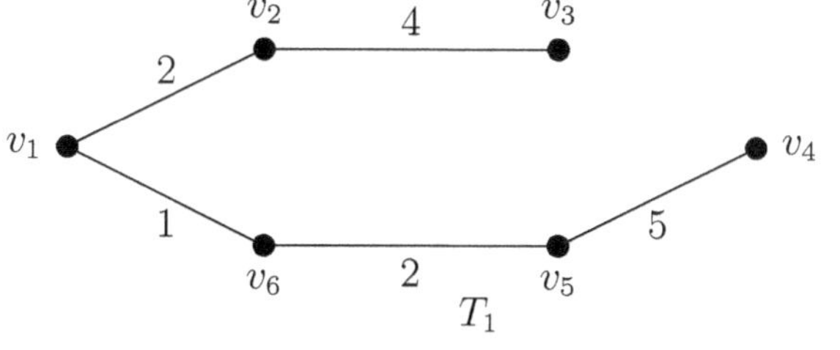

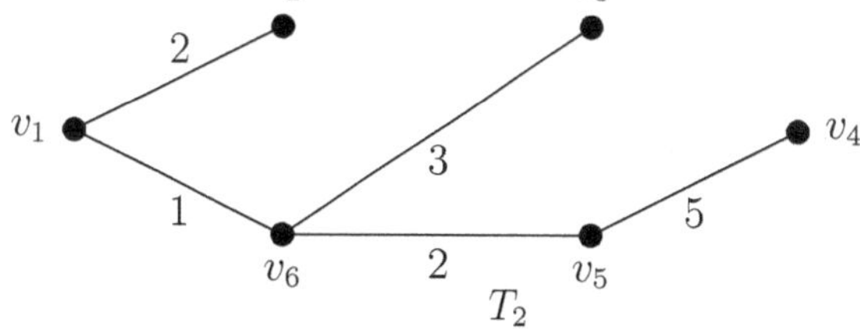

$w(T_1) = 4 + 2 + 1 + 2 + 5 = 14$ $w(T_2) = 2 + 1 + 3 + 2 + 5 = 13.$

4.3.1   Shortest spanning tree, Kruskal's and Prims algorithm

A connected weighted graph G can have more than one spanning trees. Among those, the tree with minimum weight is called as shortest spanning tree or minimal spanning tree.

If the graph is complicated, then it is difficult to list all spanning trees of the graph. So we now see some algorithms to find shortest spanning tree.

In this section, we represent two algorithms for constructing shortest spanning tree.

Kruskal's algorithm:

Let G be a weighted connected graph. First we arrange the edges of G in non-decreasing order of their weights. Next, we choose an edge with the smallest weight and which is not a loop. Successively, we add edges with minimum weight that do not form a circuit with those edges already chosen. We stop the procedure after selection of $n - 1$ edges.

Kruskal's Algorithm (Stepwise):

Let G be a connected weighted graph on n vertices.

Step 1: Select an edge e_1 of G such that e_1 is not a loop and weight of e_1 is minimum.

Step 2: If the edges e_1, e_2...e_k have been selected, then select an edge e_{k+1} from remaining edges such that

    i) weight of e_{k+1} is as small as possible
    ii) e_{k+1} does not form a circuit with previously chosen edges.

Step 3: If $n - 1$ edges are selected then stop, otherwise repeat step 2.

Example 4.9. *Using Kruskal's algorithm, find shortest spanning tree for the following graph.*

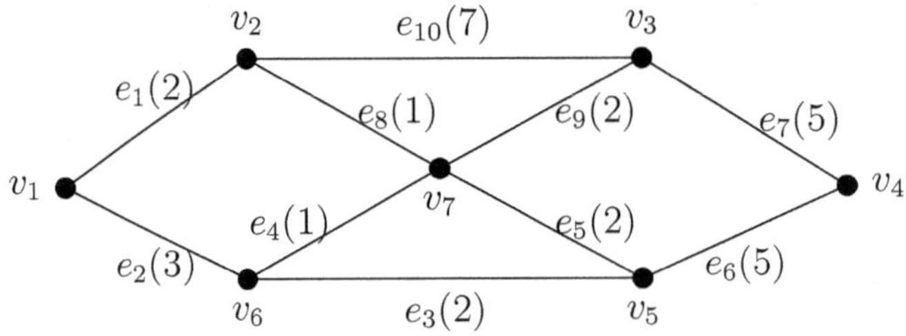

Figure 4.10

Solution: First we list the edges and their weights.

Edges	e_1	e_2	e_3	e_4	e_5	e_6	e_7	e_8	e_9	e_{10}
Weights	2	3	2	1	2	5	5	1	2	7

Now we arrange the edges in non-decreasing order of their weights.

Edges	e_4	e_8	e_1	e_3	e_5	e_9	e_2	e_6	e_7	e_{10}
Weights	1	1	2	2	2	2	3	5	5	7

Since this graph contains 7 vertices, we have to select 6 edges.

Step 1: We select edge e_4 with minimum weight.

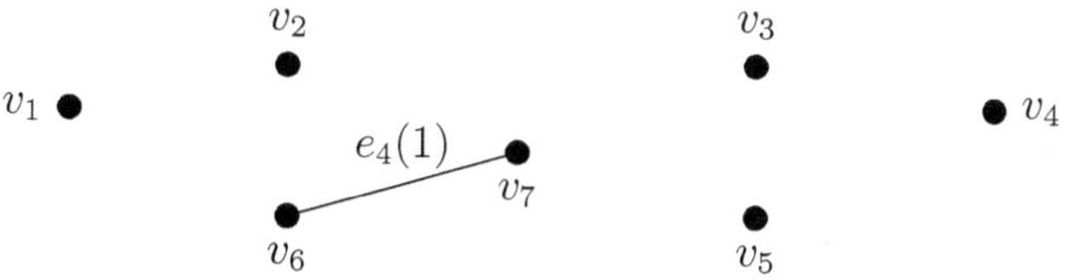

Step 2: Now we select edge e_8 which does not form circuit with previous selected edge.

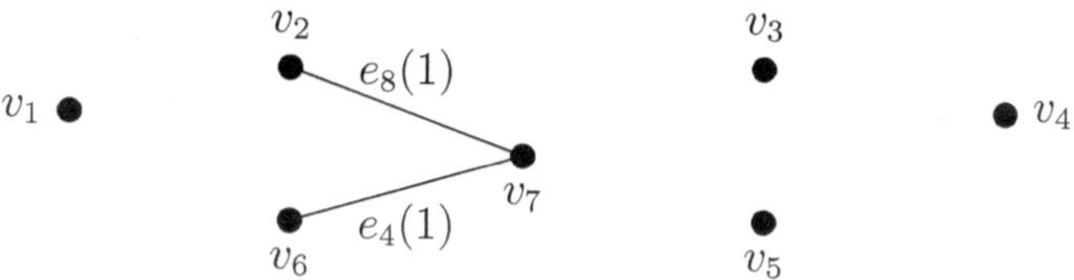

Step 3: Now we select edge e_1, because it does not form circuit with previous edges.

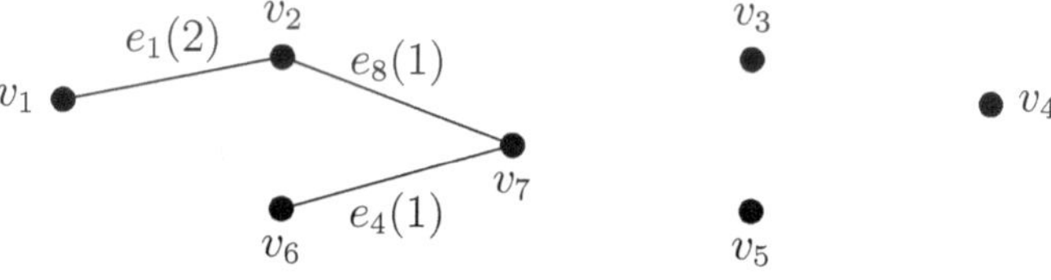

Step 4: Our next selection is edge e_3.

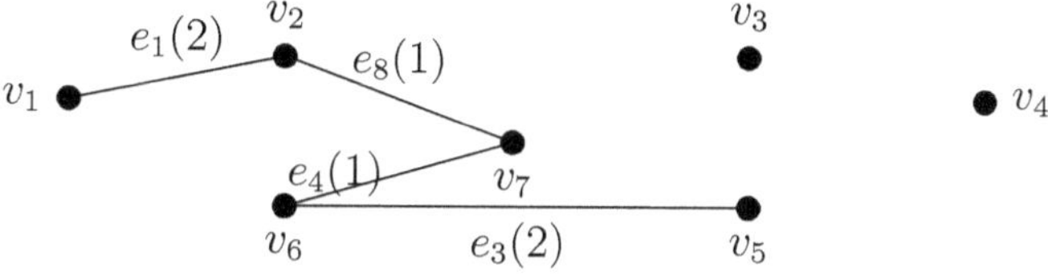

Step 5: Now next edge with minimum weight is e_5. But edge e_5 will form circuit $v_7 e_4 v_6 e_3 v_5 e_5 v_7$. So we do not select e_5. So we go for the next edge e_9. Note that edge e_9 does not form circuit.

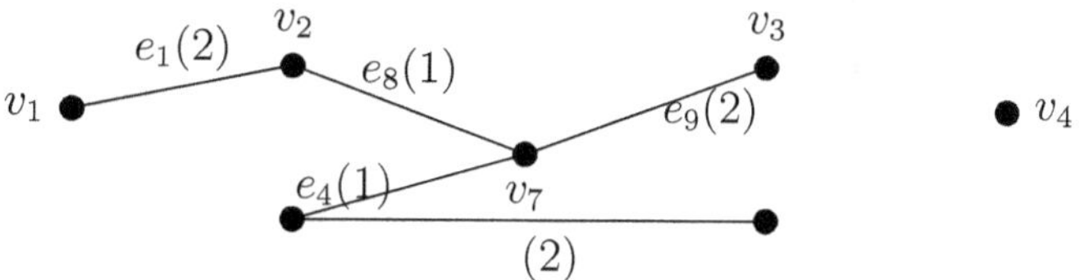

Step 6: Now we reject next edge e_2, since it forms a circuit $v_1 e_1 v_2 e_8 v_7 e_4 v_6 e_2 v_1$. We select next edge e_6 because it does not form a circuit.

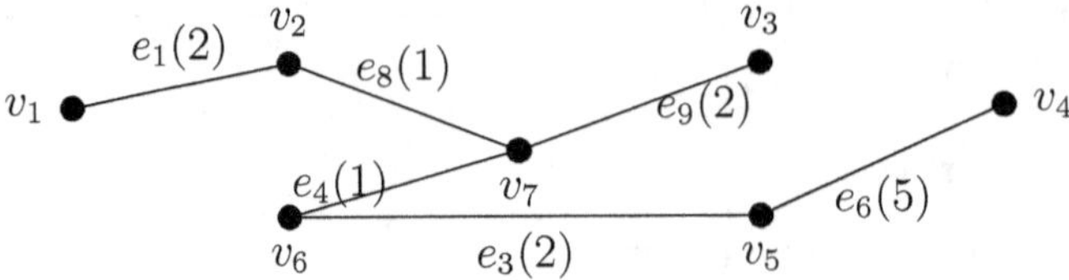

We have selected 6 edges. Hence the algorithm stops here. The graph formed by the selected edges is the shortest spanning tree. The weight of this shortest spanning tree is $1 + 1 + 2 + 2 + 2 + 5 = 13$.

Example 4.10. *Find shortest spanning tree for the following graph using Kruskal's algorithm.*

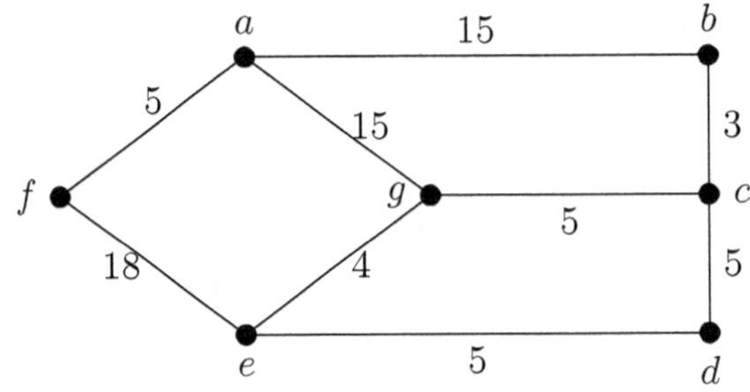

Figure 4.11

Solution: First we redraw the graph by assigning names to the edges.

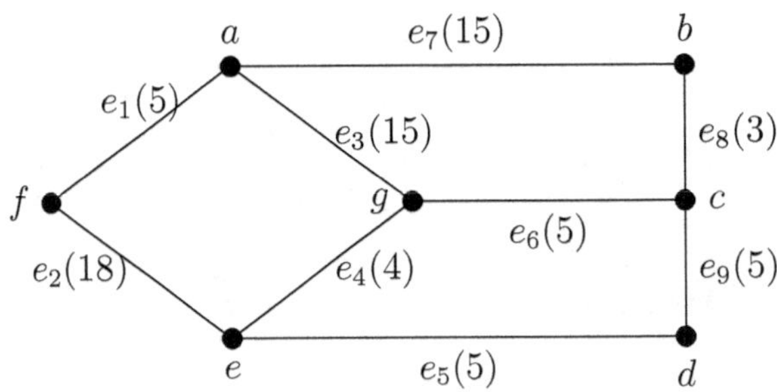

The table of edges along with corresponding weights is shown below.

Edges	e_1	e_2	e_3	e_4	e_5	e_6	e_7	e_8	e_9
Weights	5	18	15	4	5	5	15	3	5

We rearrange the edges with non-decreasing weights.

Edges	e_8	e_4	e_1	e_9	e_6	e_5	e_3	e_7	e_2

Step 1: Choose edge e_8.

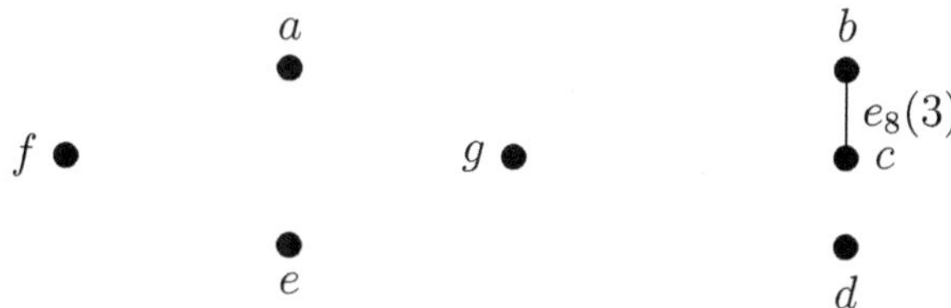

Step 2: Choose edge e_4.

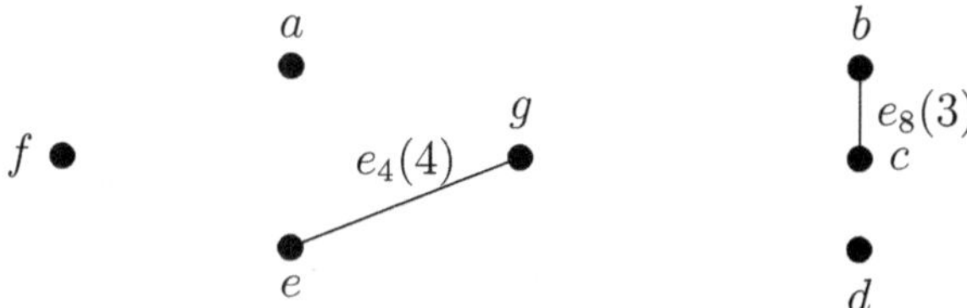

Step 3: Choose edge e_1.

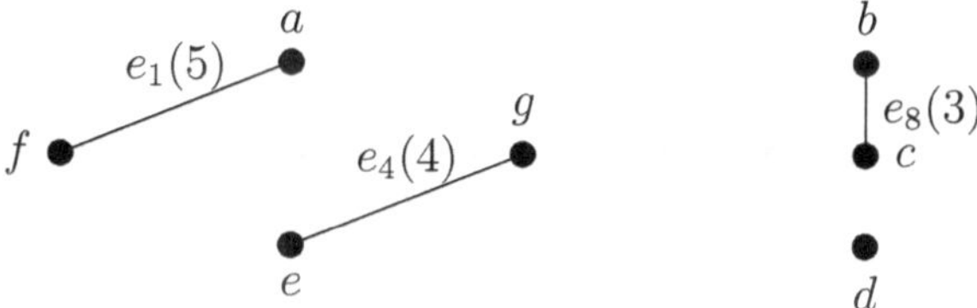

Step 4: Choose edge e_9.

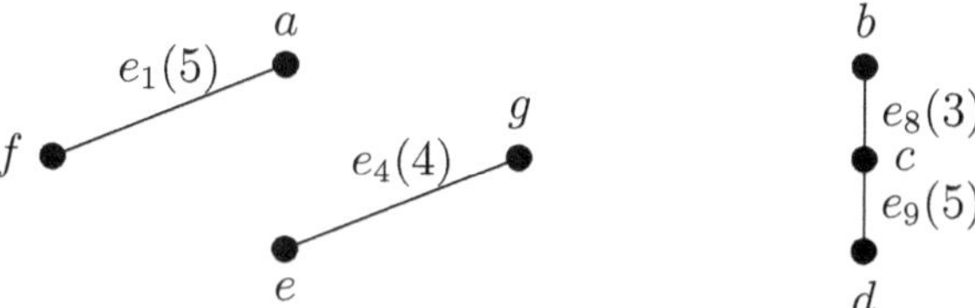

Step 5: Choose edge e_6.

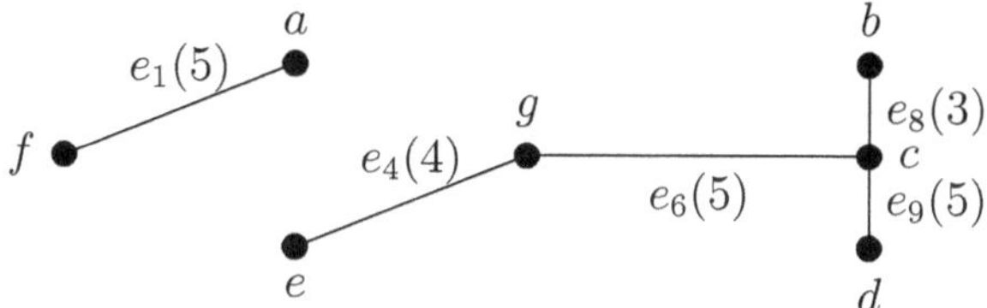

Step 6: Reject e_5, since it forms a circuit. So select edge e_3.

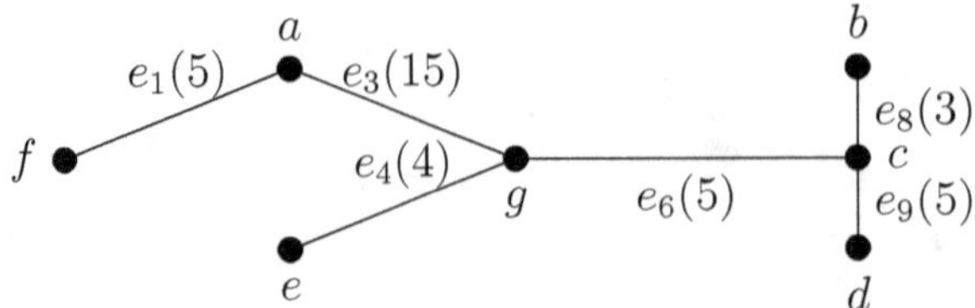

Since the graph is with 7 vertices and we have selected 6 edges, so we stop the procedure.

Weight of the shortest spanning tree is $3 + 5 + 5 + 4 + 15 + 5 = 37$.

Prims algorithm:

This algorithm is used for finding the shortest spanning tree of the given connected graph. In this we start with any vertex say v_1 of the given connected graph. Next we choose an edge of the smallest weight say e_1 which is incident on v_1 and which is not a self loop. Suppose that the other end vertex of this selected edge is v_2. Next we choose an edge of minimum weight such that its one end vertex is v_1 or v_2 and other end vertex is different from v_1 and v_2. We repeat the procedure of choosing an edge of minimum weight such that its one end vertex is one of the already chosen vertices and the other end vertex is different from the chosen vertices. We stop the procedure after selection of $n - 1$ edges.

Prims algorithm(Step wise).

Let G be a connected graph on n vertices.

Step 1: Select any vertex say v_1 of G.

Step 2: Select an edge e_1 of G such that e_1 is not a self loop and e_1 is having minimum weight among all edges incident on v_1.

Step 3: If $\{e_1, e_2...e_i\}$ are the selected edges with end vertices $\{v_1, v_2...v_{i+1}\}$ then we select an edge e_{i+1} of minimum weight such that one vertex of e_{i+1} belongs to $\{v_1, v_2...v_{i+1}\}$ and the other end vertex does not belong to $\{v_1, v_2...v_{i+1}\}$.

Step 4: If $n - 1$ edges are selected then stop otherwise repeat step 3.

Example 4.11. *Use Prim's algorithm to find a minimal spanning tree in the following graph G.*

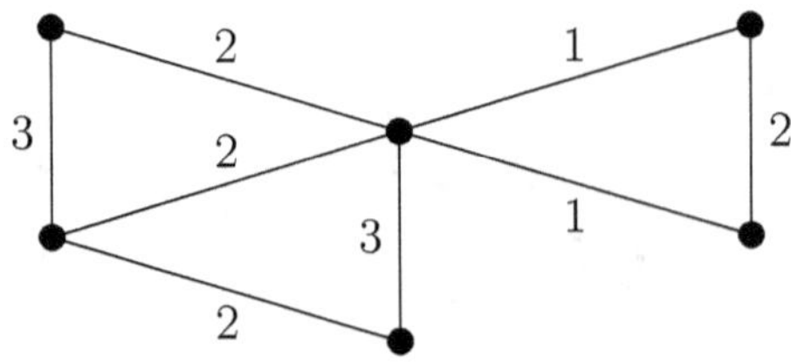

Figure 4.12

Solution: We first give names to the vertices and edges.

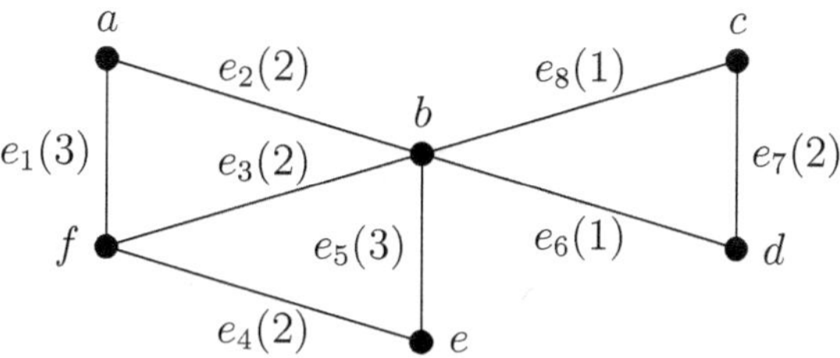

Step 1: Suppose we start with vertex b.

Step 2: Now there are 5 edges incident on b. Among those edges e_6 and e_8 both are of minimum weights. Suppose we choose edge e_8.

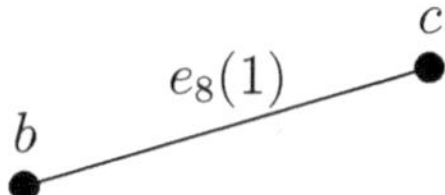

Step 3: Now the edges e_2, e_3, e_5, e_6 incident on b have weights 2, 2, 3, 1 respectively. Also the edge e_7 incident on c has weight 2, so we choose edge e_6 with minimum weight 1.

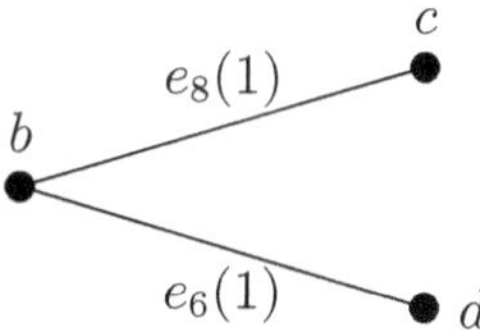

Step 4: Now edge e_7 has both end vertices in the already chosen vertices c and d. So we do not select e_7. We see only those edges incident on b. Among the edges e_2, e_3 and e_5, the edges e_2 and e_3 have smallest weight 2. So we select e_2.

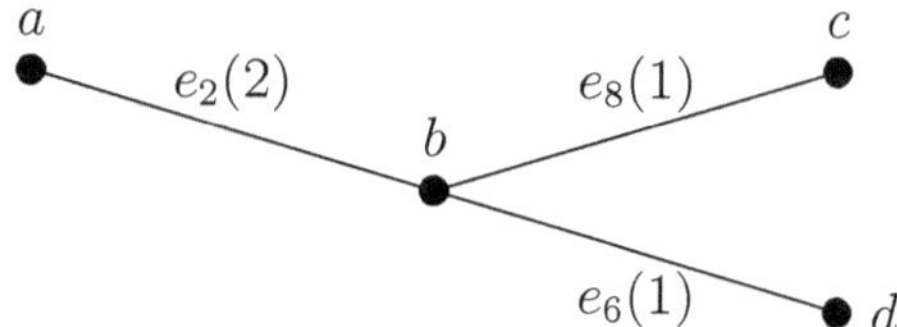

Step 5: Now edge e_1 is incident on vertex a and edges e_3 and e_5 are incident on b. We choose edge e_3 with smallest weight 2.

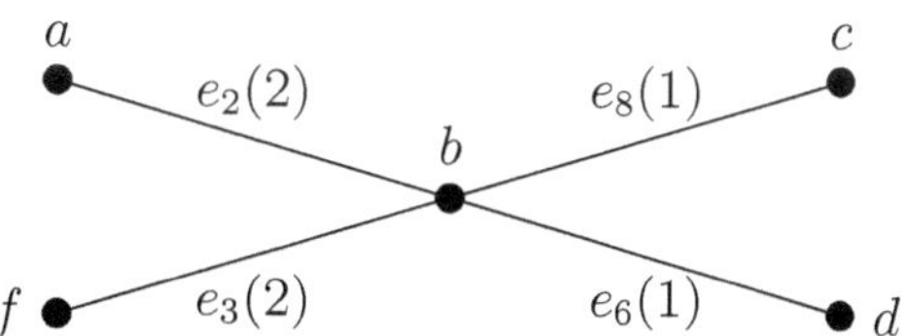

Step 6: Now edge e_1 cannot be selected since its both end vertices a and f belong to the chosen set.

e_4.

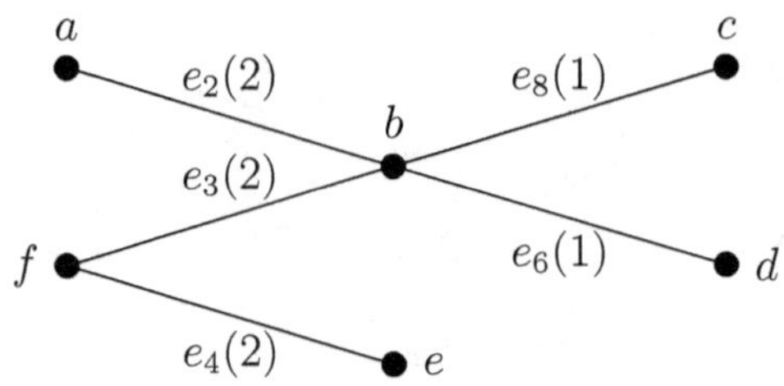

Since the given graph is on 6 vertices, we stop the procedure here since we have selected 5 edges.
Weight of the minimal spanning tree $= 1 + 1 + 2 + 2 + 2 = 8$.

Example 4.12. *Find the shortest spanning tree for the following graph using Prim's algorithm.*

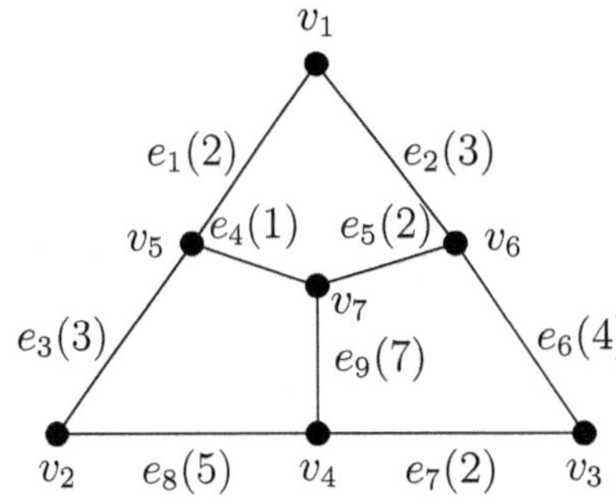

Figure 4.13

Solution:

Step 1:

Step 2:

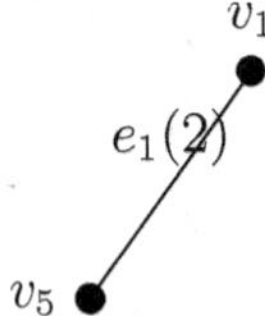

Step 3:

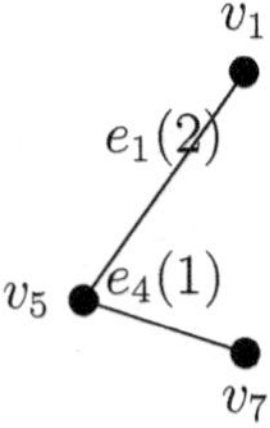

Step 4:

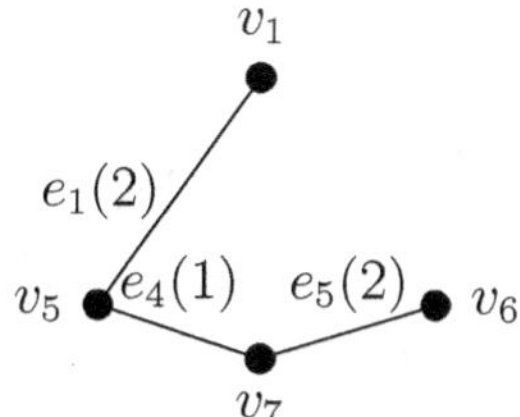

Step 5:

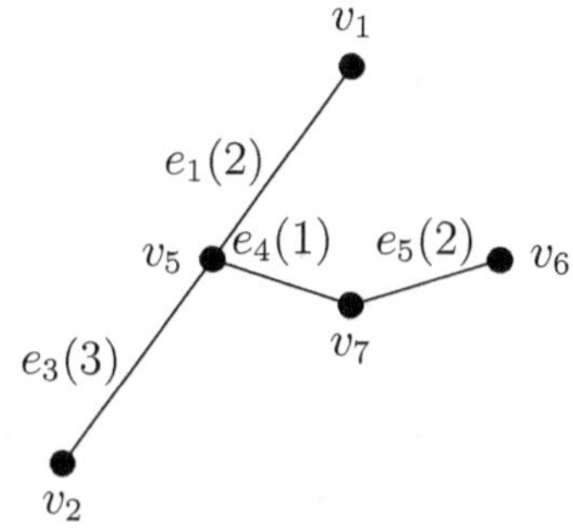

Step 6:

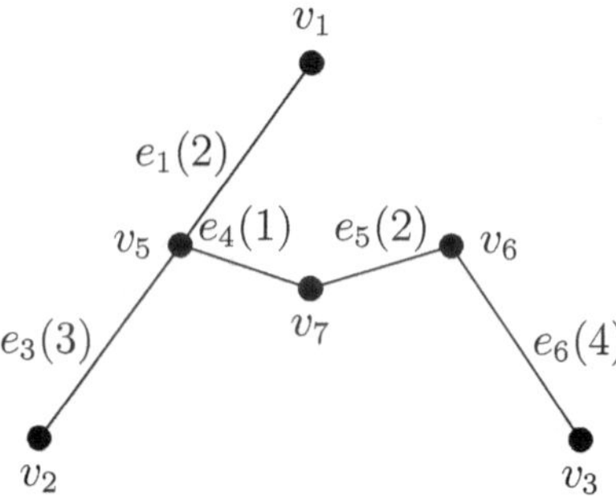

Step 7:

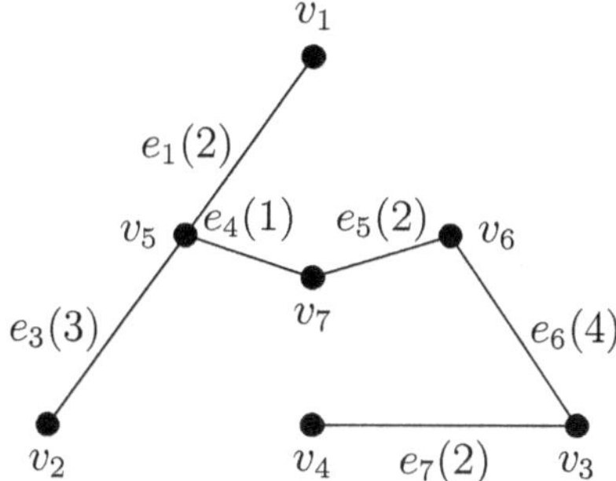

Hence weight of the shortest spanning tree $= 2 + 3 + 1 + 2 + 4 + 2 = 14$.

Example 4.13. *Find the minimal spanning tree of the weighted graph using Prim's algorithm.*

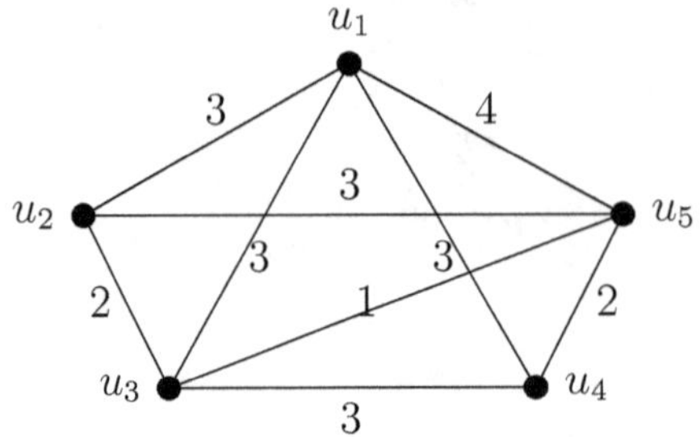

Figure 4.14

Solution:

Step 1: Suppose we start with u_4.

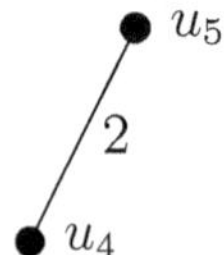

Step 2: Among the three edges incident on u_4, we select edge with minimum weight 2.

Step 3: Edges incident on u_4 and u_5 have weights 4, 3, 1, 3 and 3. So we choose the edge with weight 1.

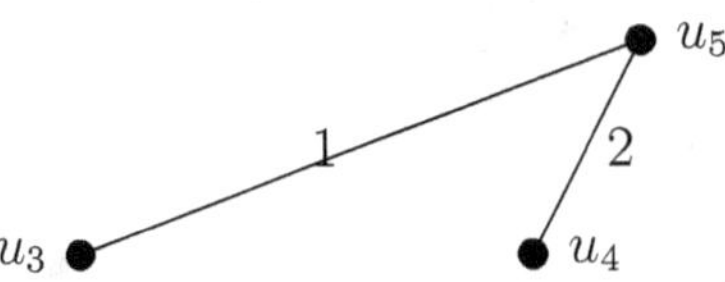

Step 4: Next we select edge with weight 2 incident on u_3.

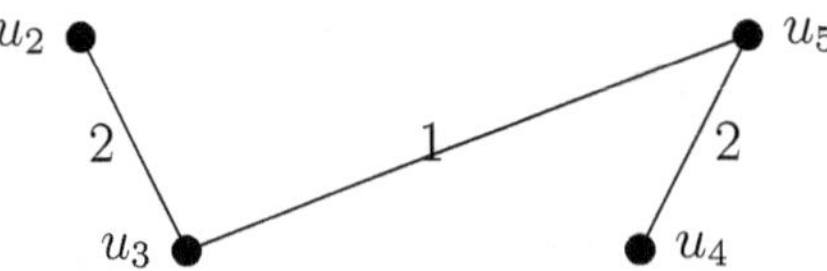

Step 5: Select edge with weight 3 incident on u_2.

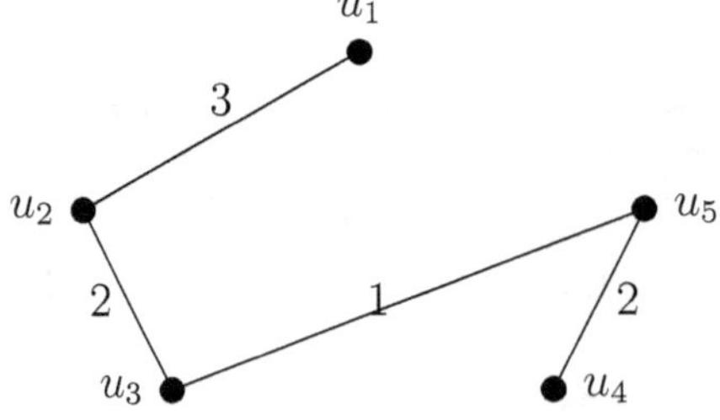

We stop the procedure, since we have selected 4 edges.

4.4   Binary tree

In this section, we will first study binary trees and then we further study its generalization that is m-ary trees.

Definition 4.8. *Rooted tree:* *A tree in which one vertex is distinguished from all other vertices is called a rooted tree.*

For example, the tree shown below is a rooted tree.

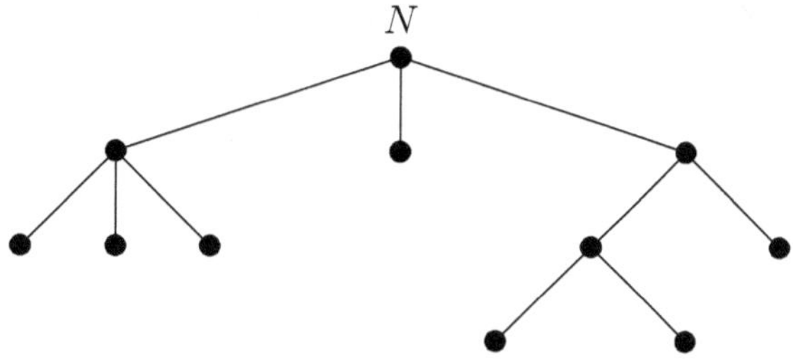

Figure 4.15

In this tree the vertex N is the root of the tree.

Definition 4.9. *Binary tree:* *A tree in which there is exactly one vertex of degree 2 and all other vertices are of degree 1 or 3 is called as binary tree.*

For example, the tree shown below is a binary tree.

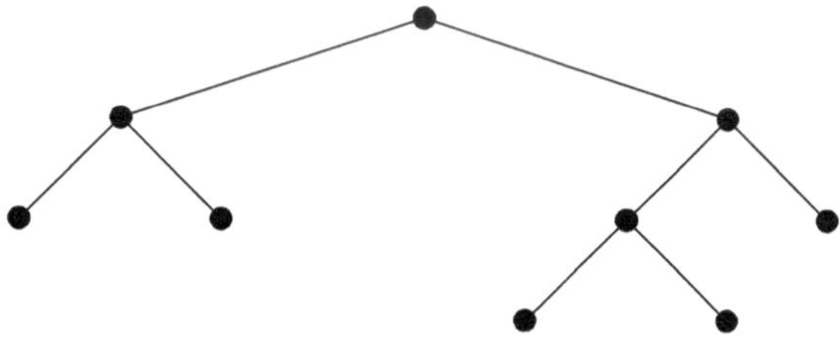

Figure 4.16

Definition 4.10. *Level of a vertex:* *Let T be a tree with root 'r'. Then the distance of the vertex 'v' from root 'r' is called as level of vertex 'v'.*

Definition 4.11. *Height of a binary tree:* *The maximum level in a binary tree is known as the height of a binary tree.*

Remark 4.2. *1) The maximum height of a binary tree with n vertices is $(n-1)/2$.*
2) The minimum height of a binary tree with n vertices is $[log_2(n+1) - 1]$, where $[x]$ denotes the smallest integer greater than or equal to x.

Example 4.14. *In the following binary tree, find*

2) level of v_5, v_9, v_{13}

3) height of the tree

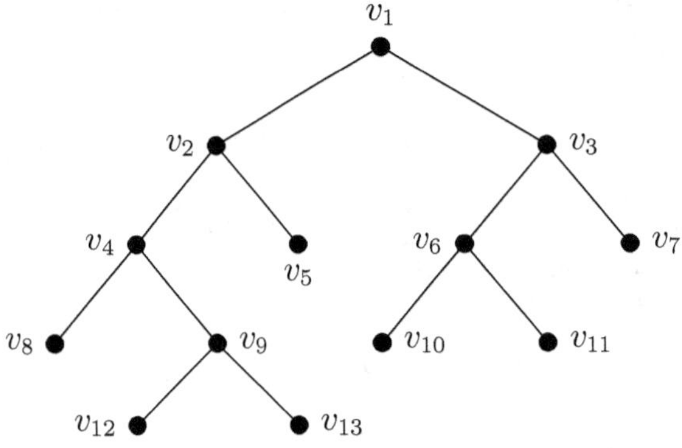

Figure 4.17

Solution:

1) root : v_1

2) level of v_5=2, level of v_9=3, level of v_{13}=4

3) height of tree = 4.

Theorem 4.7. *The number of vertices in a binary tree is always odd.*

Proof. Let T be a binary tree on n vertices. Then there is only one vertex of even degree that is of degree 2 and all remaining $n-1$ vertices are of degree 1 or 3. That is there are $n-1$ vertices of odd degree. But we know that the number of odd degree vertices is always even. Hence $n-1$ is even. So we get that n is odd. $\square$

Theorem 4.8. *The number of pendant vertices in a binary tree on n vertices is $\dfrac{n+1}{2}$ and number of non-pendant vertices is $\dfrac{n-1}{2}$.*

Proof. Let T be a binary tree on n vertices and let p be the number of pendant vertices. Hence there are $n-p-1$ vertices of degree 3 and only one vertex of degree 2. Now total number of edges in T is $n-1$. So by Handshaking theorem,

Total degree $= 2(n-1)$

That is, $1(2) + p(1) + 3(n-p-1) = 2n-2$

That is, $3n - 2p - 1 = 2n - 2$

That is, $p = \dfrac{n+1}{2}$.

Therefore the number of non-pendant vertices $= n - p = n - (n+1)/2 = \dfrac{n-1}{2}$. $\square$

Example 4.15. *Does there exits a binary tree with 2020 vertices? Justify your answer.*

Solution: Number of vertices in a binary tree is always odd. But 2020 is an even number. Hence

Example 4.16. *Determine maximum and minimum height of a binary tree with 11 vertices. Also draw such trees.*

Solution: Maximum height $= (n-1)/2 = (11-1)/2 = 5$

Minimum height $= [log_2(n+1) - 1] = [log_2(12) - 1] = [2.58] = 3$.

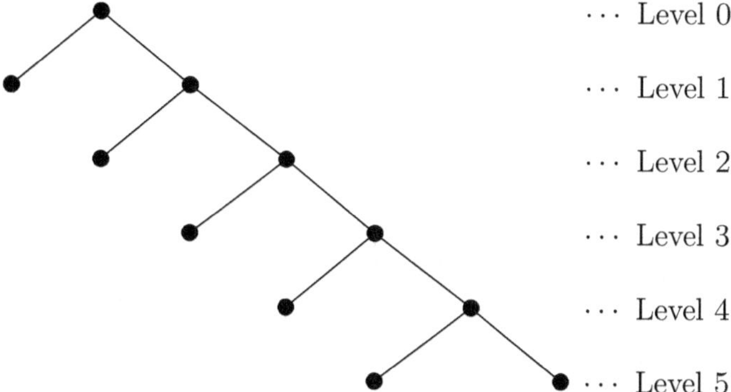

Binary tree with maximum height

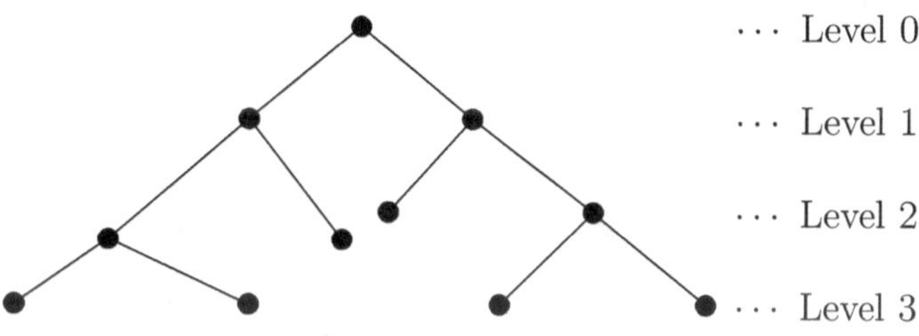

Binary tree with minimum height

Figure 4.18

Example 4.17. *Draw all non-isomorphic binary trees of height* 2.

Solution:

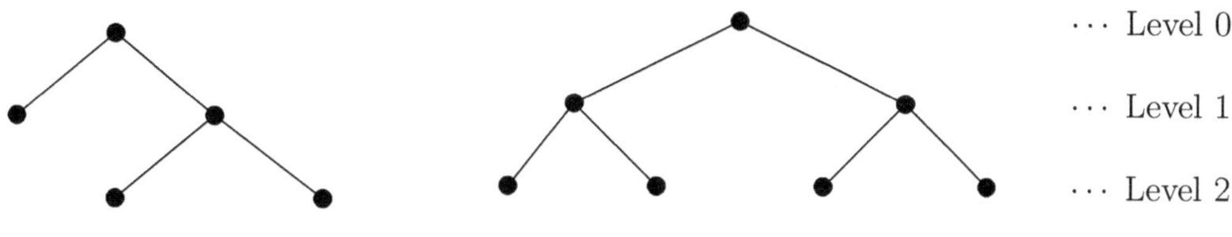

Figure 4.19

4.5   m-ary trees:

In a rooted tree, we can assign directions to each edge as follows: Since there is unique path from root to each vertex of the graph, we direct each edge away from the root. We usually draw a rooted tree with its root at the top of the graph. For example,

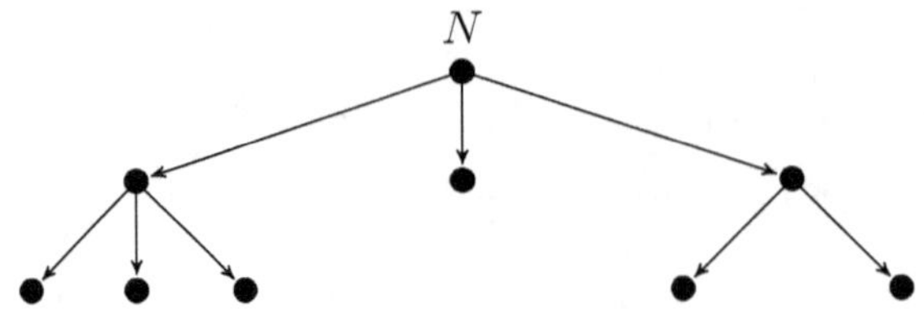

Figure 4.20

Since the choice of the root clearly determines the direction, we can omit the arrows indicating directions of the edges.

Definition 4.12. *If in a rooted tree, there is a directed edge from vertex u to v, then we say that 'u is parent of v' and 'v is child of u'. A vertex having no child is called as leaf. Vertices that have children are called internal vertices.*

Example 4.18. *Consider the following rooted tree. Write internal vertices and leaves of the tree. Find children of v_2 and parent of v_{12}. Draw the subtree rooted at vertex v_3.*

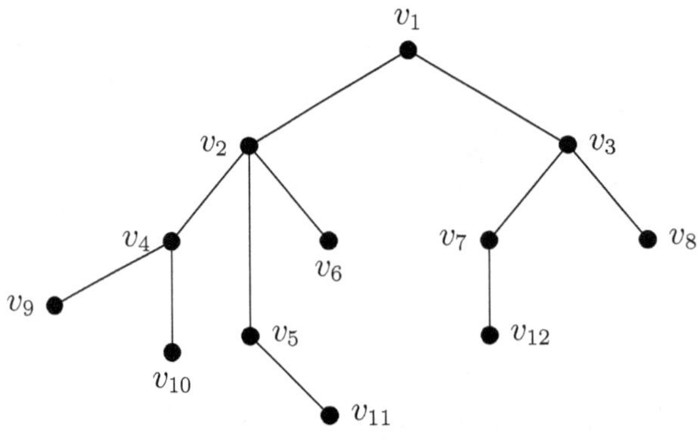

Figure 4.21

Solution:

Internal vertices: v_1 ,v_2 ,v_3 ,v_4 ,v_5 , v_7 Leaves: v_9, v_{10} ,v_{11} ,v_{12} ,v_8, v_6

Children of v_2: v_4 ,v_5 ,v_6 Parent of v_{12}: v_7

Subtree at vertex v_3:

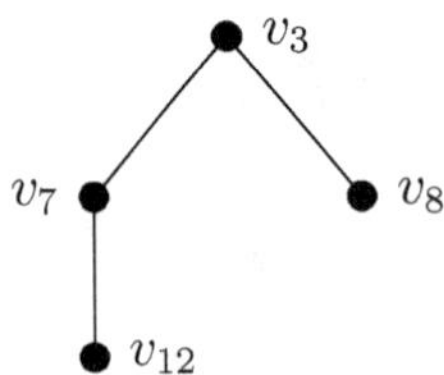

Definition 4.13. *A rooted tree is called m-ary tree if every internal vertex has no more than m*

Illustrative example: Consider the trees given below:

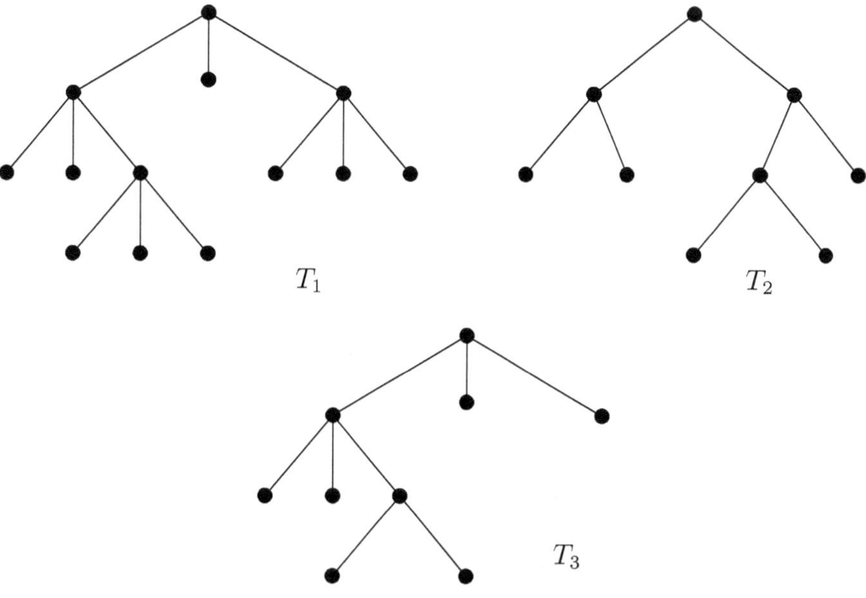

Figure 4.22

In tree T_1, each internal vertex has 3 children. Hence T_1 is a full 3-ary tree. In tree T_2, each internal vertex has exactly two children. So T_2 is a full 2-ary tree. A full 2-ary tree is nothing but a binary tree which we have studied. In tree T_3, one internal vertex has 3 children while other two has 2 children. So it is not a full m-ary tree for any m.If the children of each internal vertex are ordered then we say that the tree is a ordered rooted tree.

For example, in a binary tree, if an internal vertex has two children, the first child is called the left child and the second child is called the right child. Also the tree rooted at the left child is called left subtree and the tree rooted at the right child is called right subtree.

Example 4.19. *What are left and right children of vertex e in the following binary tree. Also draw left and right subtrees of vertex c.*

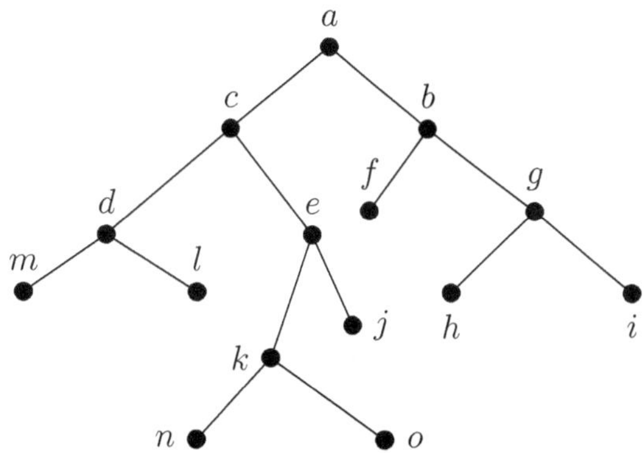

Figure 4.23

Solution: Left child of e: k

Right child of $e : j$

Subtrees of C:

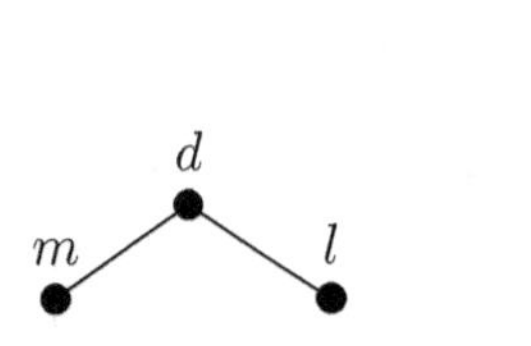

Left subtree of vertex c   Right subtree of vertex c

Properties of m-ary tree:

Theorem 4.9. *A full m-ary tree with i internal vertices has $n = m.i + 1$ vertices.*

Proof. In m-ary tree, each of the i internal vertices has m children. Hence there are $m.i$ vertices. But the root is not a child of any vertex. Therefore total number of vertices is $m.i + 1$ $\qquad\square$

Theorem 4.10. *A full m-ary tree with*
i) n vertices has $i = \dfrac{n-1}{m}$ internal vertices and $l = \dfrac{(m-1)n+1}{m}$ leaves.
ii) i internal vertices has $n = m.i + 1$ vertices and $l = (m-1)i + 1$ leaves.
iii) l leaves has $n = \dfrac{(m-l-1)}{(m-1)}$ vertices and $i = \dfrac{l-1}{m-1}$ internal vertices.

Proof. i) Suppose a m-ary tree has n vertices. By previous theorem, if it contains i internal vertices then $n = m.i + 1$. Therefore $i = \dfrac{n-1}{m}$

Also every vertex is either a leaf or an internal vertex. Hence $n = l + i$

This gives that number of leaves is $l = n - i$
$$= n - \frac{n-1}{m}$$
$$= \frac{(m-1)n+1}{m}.$$

ii) We have already proved that $n = m.i + 1$.

Now $l = n - i = m.i + 1 - i = (m-1)i + 1$

iii) Substituting $n = l + i$ in $n = m.i + 1$, we get

$$l + i = m.i + 1$$
$$(m-1)i = l - 1$$
$$i = \frac{l-1}{m-1}.$$

$$l - 1 \qquad\qquad ml - 1$$

Theorem 4.11. *There are almost m^h leaves in a m-ary tree of height h.*

Proof. We prove this by induction on height h. If height of a binary tree is 1, then it consists of a root with no more than m children. Hence there are no more than $m^1 = m$ leaves. So result holds for h=1.

Suppose the result holds for m-ary tree of height less than k. Let T be a m-ary tree with height k. After deletion of the root, we get at most m subtrees. Height of each subtree is less than or equal to $k - 1$. So by induction hypothesis, each of these subtrees have almost m^{k-1} leaves. Since there are almost m subtrees, the maximum number of leaves is $m.(m^{k-1})$ that is m^k leaves as desired. Hence by induction, the result holds for all h. $\qquad\square$

4.6   Tree Traversal

Ordered rooted trees are often used to store information. We need procedures for visiting each vertex of an ordered rooted tree to access data.In this section we will see several algorithms for visiting all vertices of an ordered rooted tree. These ordered rooted trees can also be used to represent various types of expressions such as arithmetic expressions involving numbers, variables and operations. The different listing of vertices of ordered rooted tree used to represent expressions are useful in the evaluation of these expressions.

Traversal algorithm: Procedures for systematically visiting every vertex of an ordered rooted tree are called traversal algorithms. We will study three such algorithms, preorder trversal, inorder traversal and postorder traversal.

Definition 4.14. *Preorder traversal:* *Let T be an ordered rooted tree with root r. If T consists only of r, then r is the preorder traversal of T. Otherwise suppose that T_1, T_2, $\cdots$, T_n are the subtrees at r from left to right in T. The preorder traversal begins by visiting r.It continues by traversing T_1 in preorder, then T_2 in preorder and so on until T_n is traversed in preorder.*

The algorithm is shown in the following figure.

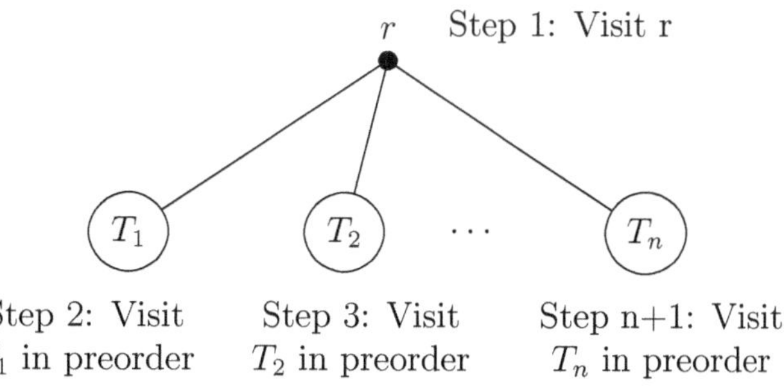

Example 4.20. *In which order does a preorder traversal visit the vertices in the following ordered rooted tree?*

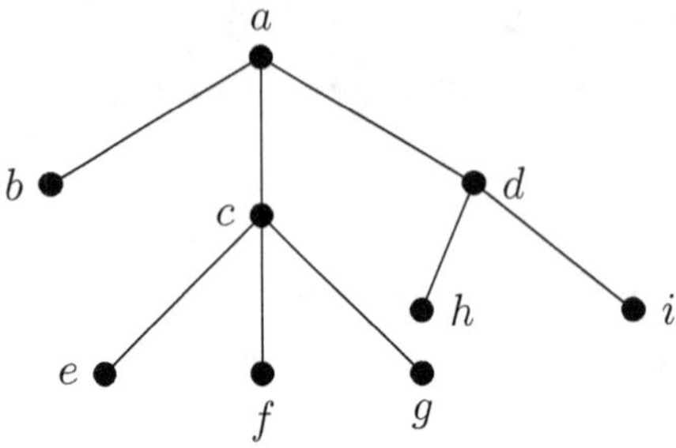

Figure 4.24

Solution: We visit root vertex a and then its subtrees from left to right in preorder.

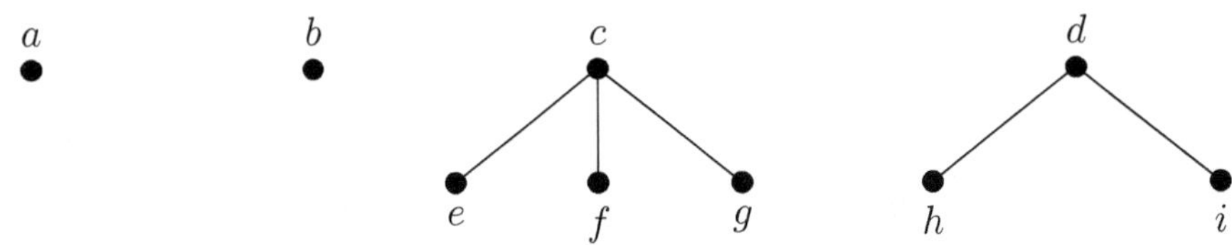

Figure 4.25

Now for visiting the subtree at c and d, we again visit the roots and then their subtrees from left to right.

Example 4.21. *In which order does a preorder traversal visit the vertices in the following ordered rooted tree?*

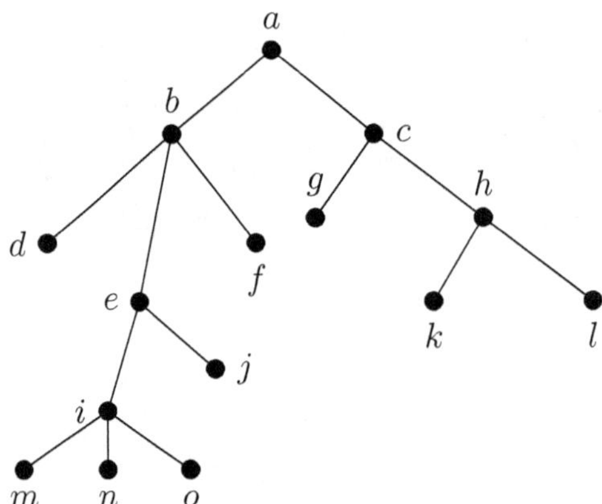

Figure 4.26

Solution:

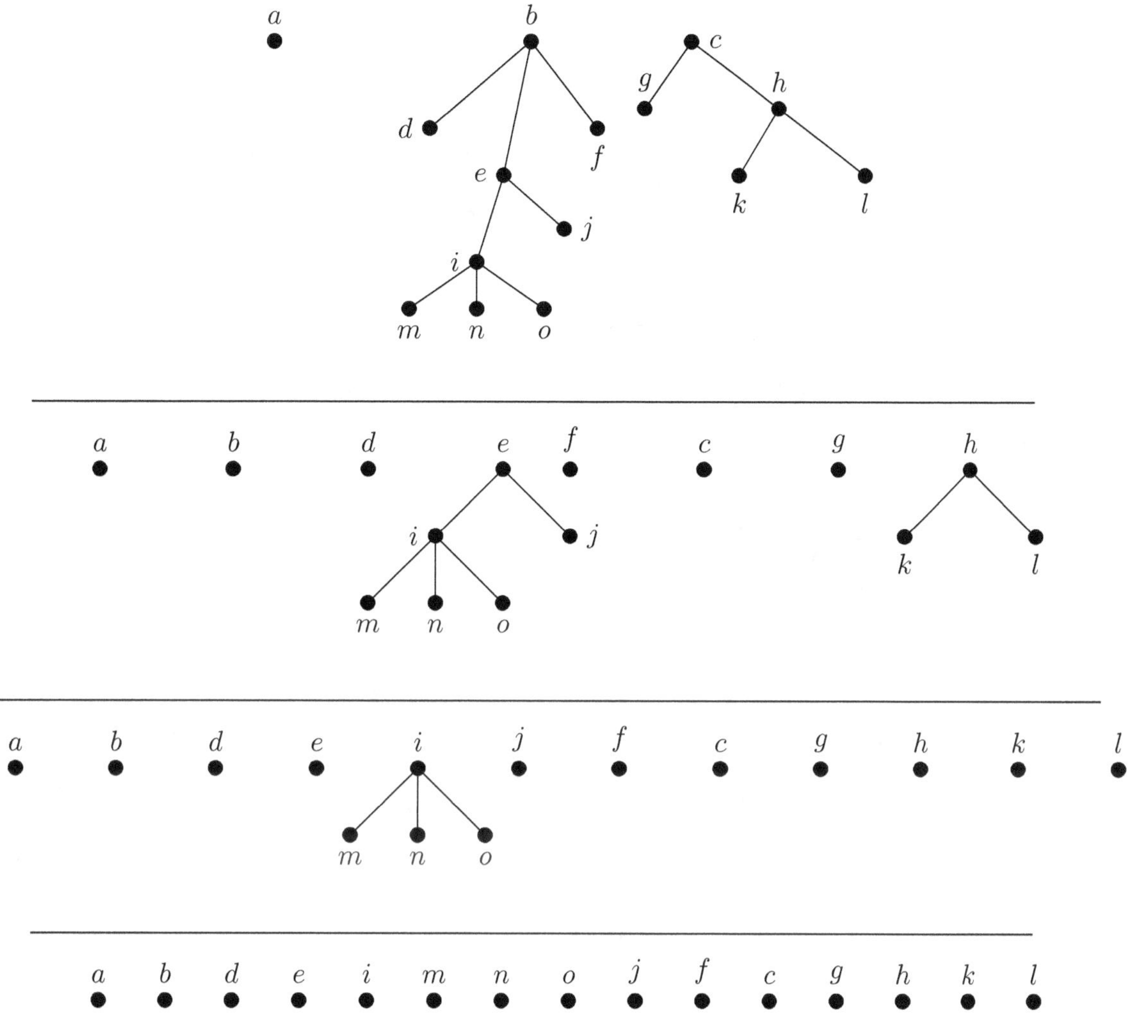

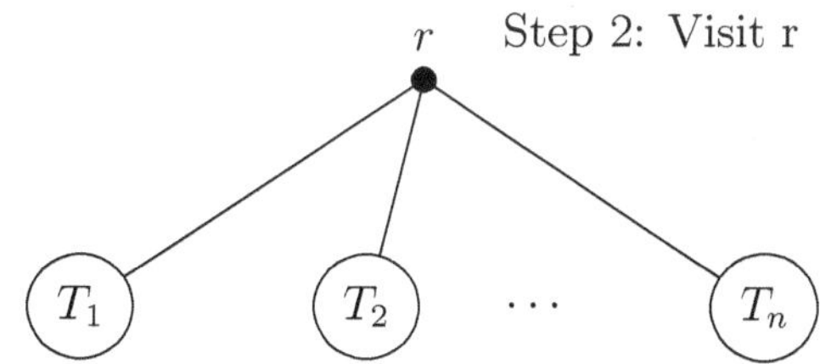

Definition 4.15. *Inorder traversal:* *Let T be an ordered rooted tree with root r. If T consists only of r, then r is the inorder traversal of T. Otherwise suppose that T_1, T_2, $\cdots$, T_n are the subtrees at r from left to right in T. The inorder traversal begins by traversing T_1 in inorder, then visiting r. It continues by traversing T_2 in inorder, then T_3 in inorder and finally T_n in inorder.*

The algorithm is shown in the following figure.

Inorder traversal

Example 4.22. *In which order does an inorder traversal visit the vertices in the following ordered rooted tree?*

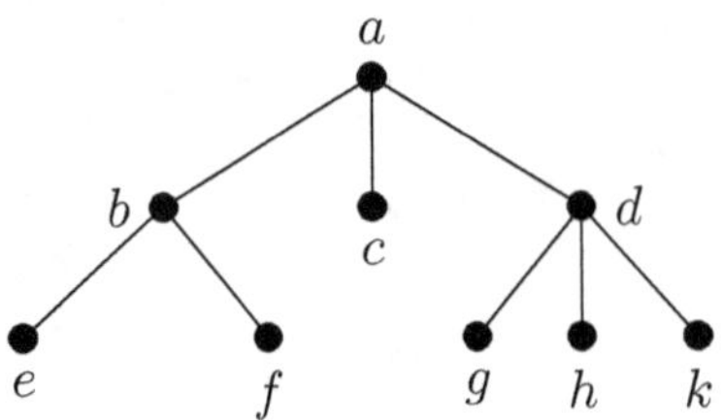

Figure 4.27

Solution: First we visit T_1 in inorder and then visit root 'a' following by remaining subtrees.

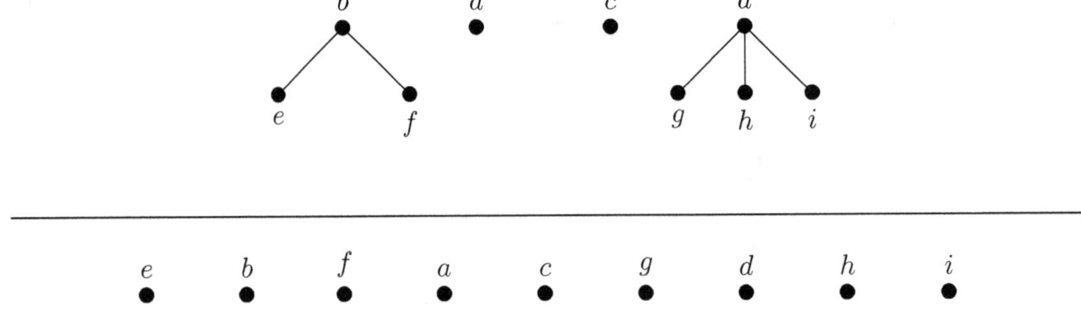

Example 4.23. *In which order does an inorder traversal visit the vertices in the following ordered rooted tree?*

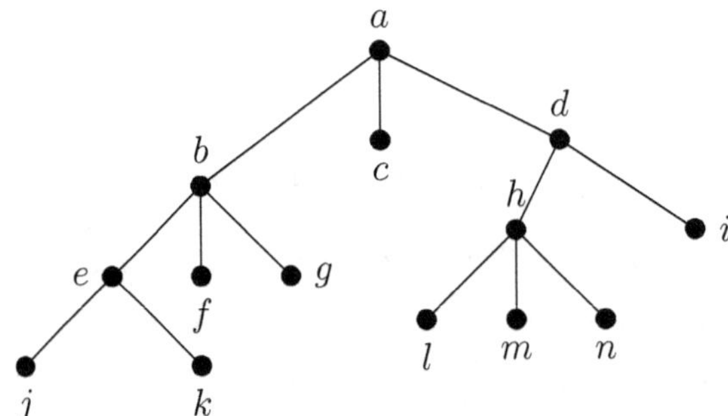

Figure 4.28

Solution:

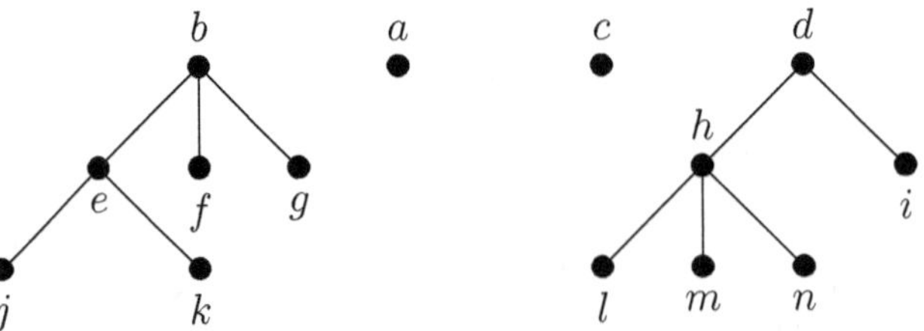

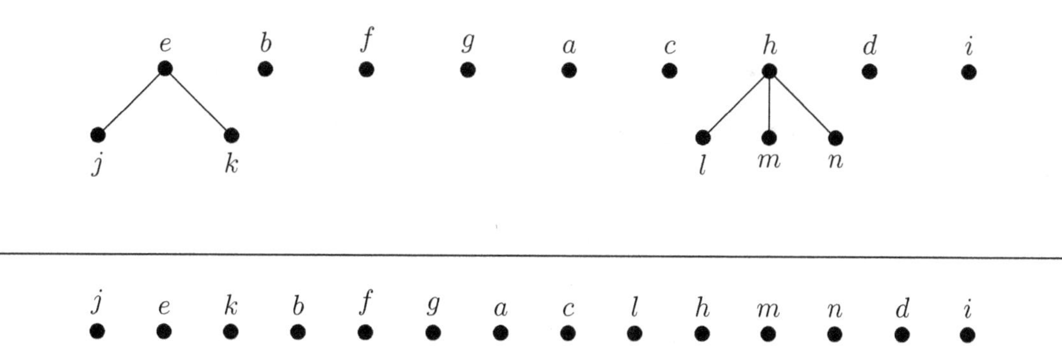

Definition 4.16. *Postorder traversal:* *Let T be an ordered rooted tree with root r. If T consists only of r, then r is the postorder traversal of T. Otherwise suppose that T_1, T_2, $\cdots$, T_n are the subtrees at r from left to right in T. The postorder traversal begins by traversing T_1 in postorder, then T_2 in postorder, ... T_n in postorder and ends by visiting r.*

The algorithm is shown in the following figure.

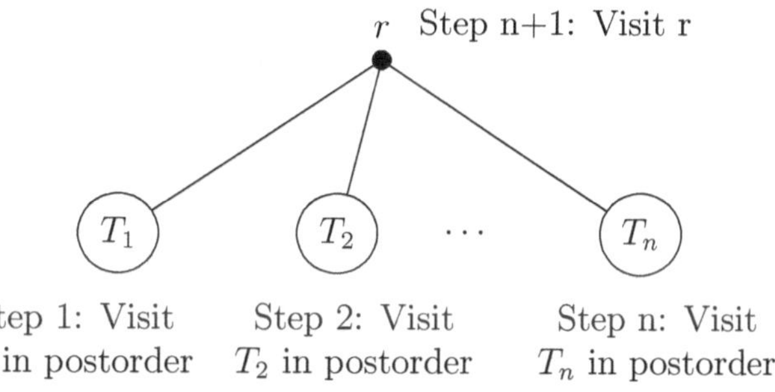

Example 4.24. *In which order does a postorder traversal visit the vertices in the following ordered rooted tree?*

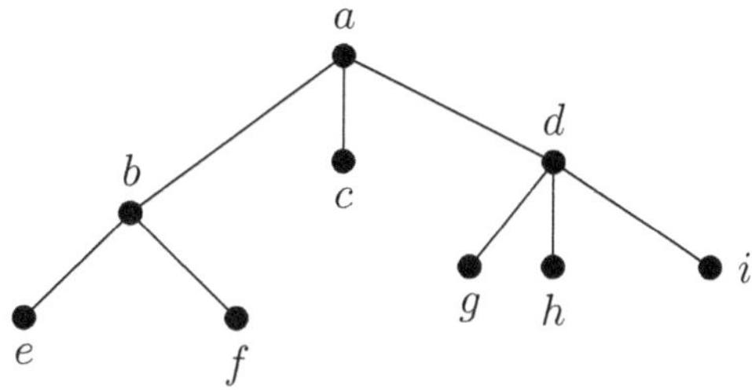

Figure 4.29

Solution: First we visit subtree at 'b', then subtree at 'c', then subtree at 'd' followed by root 'a'.

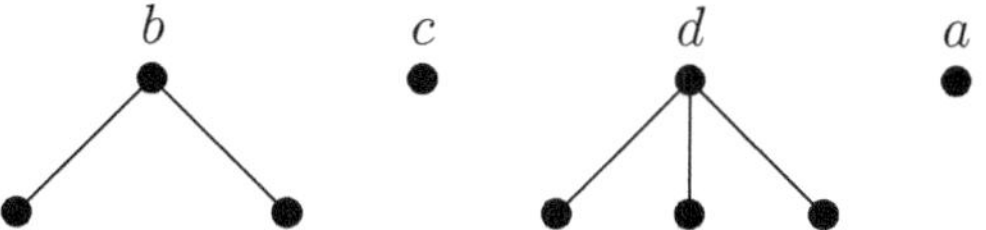

In second step we traverse subtrees at b and d in postorder.

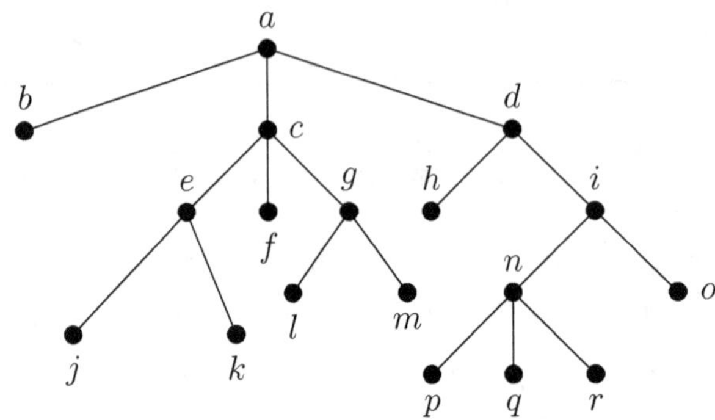

Example 4.25. *In which order does a postorder traversal visit the vertices in the following ordered rooted tree?*

Figure 4.30

Solution:

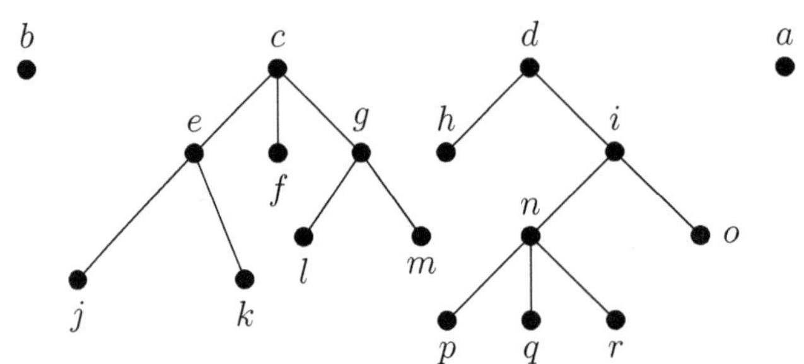

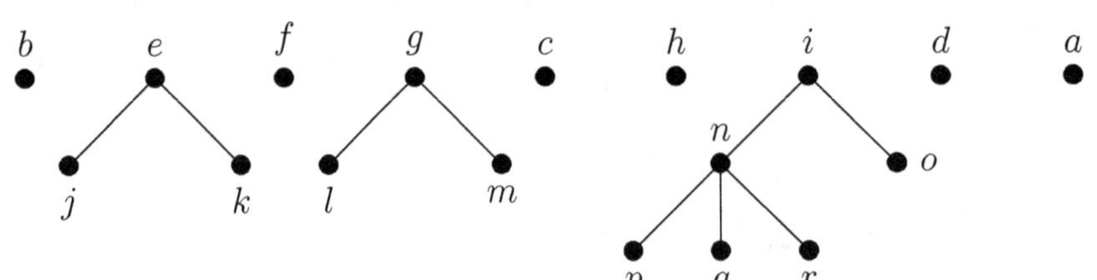

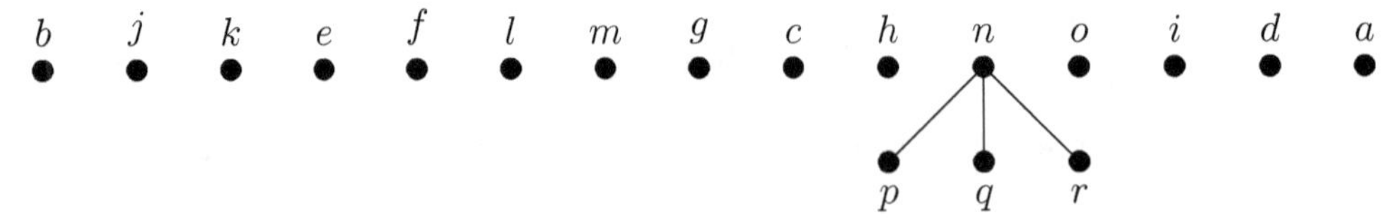

4.6.1   Infix, Prefix and Postfix notation

We can represent complicated expressions such as compound propositions, combinations of sets and arithmetic expressions using ordered rooted trees. For example if the arithmetic expression involved the operators $+$(addition), $-$(Subtraction), $*$(multiplication), $/$ (division), and $\uparrow$(exponention), then the internal vertices represent operations and the leaves represent the variables or numbers. Each operation operates on its left and right subtrees (in that order).

Example 4.26. *Find the ordered rooted tree that represent the expression $xy + (y + 3)^2$?*

Solution:

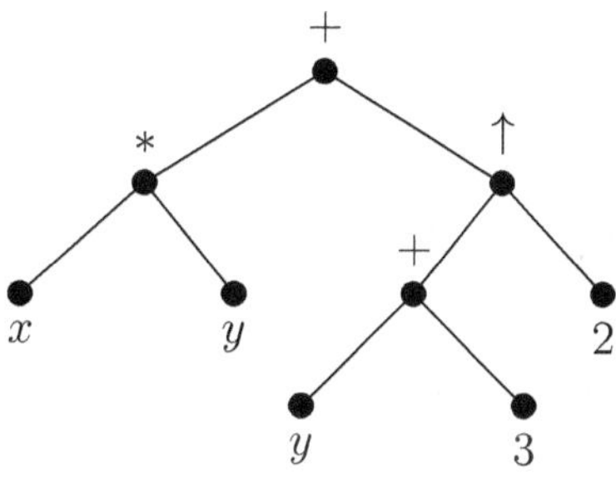

Figure 4.31

Prefix form or Polish notation: When we traverse the rooted tree representing the arithmetic expression in preorder, we obtain the prefix form of an expression. Expressions written in prefix form are said to be in Polish notation.

Example 4.27. *What is the prefix form of the following expression: $\dfrac{x + y}{3} - (x - 3)^4$.*

Solution: First we find the ordered rooted tree representing this expression.

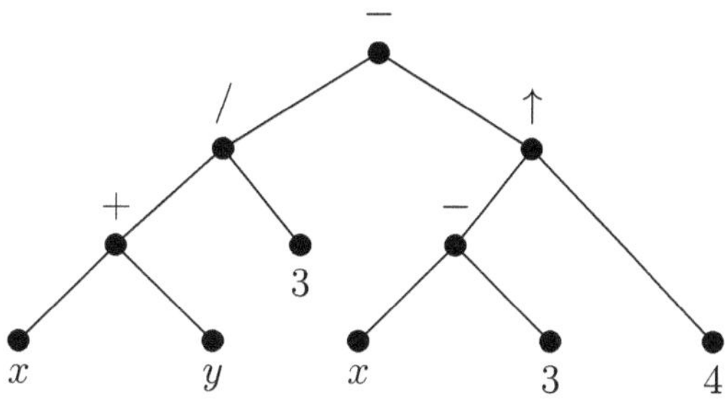

Figure 4.32

We traverse this tree in preorder to get the following prefix form.

$$- \ / \ + \ x \ y \ 3 \ \uparrow \ - \ x \ 3 \ 4$$

Postfix form or Reverse Polish notation: If we traverse the ordered rooted tree representing the arithmetic expression in postorder, we obtain the postfix form of the expression. Expressions

Example 4.28. *Write the postfix form of the following expression:* $\dfrac{2x + y}{(x + y)^3}$.

Solution: The ordered rooted tree representing this expression is

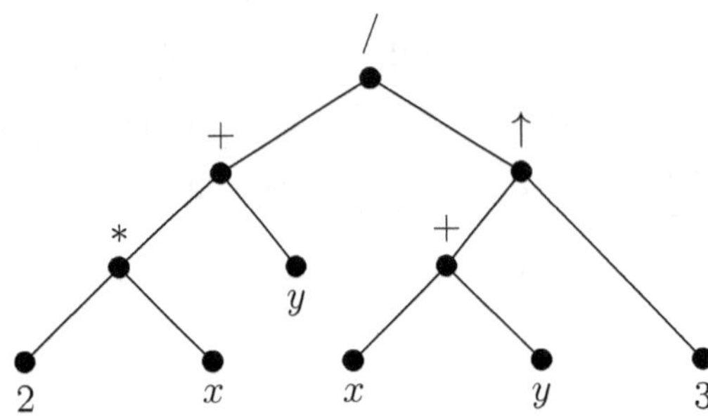

Figure 4.33

We traverse this tree in postorder to get the following postfix form.

$$2\ x\ *\ y\ +\ x\ y\ +\ 3\ \uparrow\ /$$

Infix form: An inorder traversal of the binary tree representing an expression produces the original expression. We have to include parentheses whenever we encounter an operation. The fully parenthesized expression obtained in this way is said to be in infix form.

Example 4.29. *Write the infix form of the expression representing the following rooted tree.*

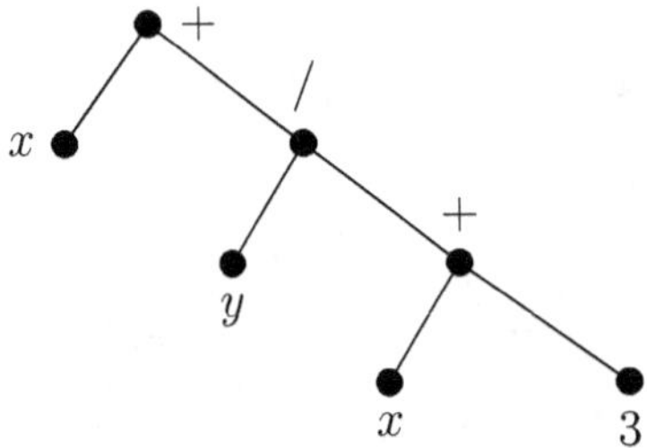

Figure 4.34

Solution: Infix form is $x\ +\ (y/(x + 3))$

Example 4.30. *Represent the expression $(x+xy)+(x/y)$ using binary tree. also write this expression in prefix, postfix and infix notation.*

Solution:

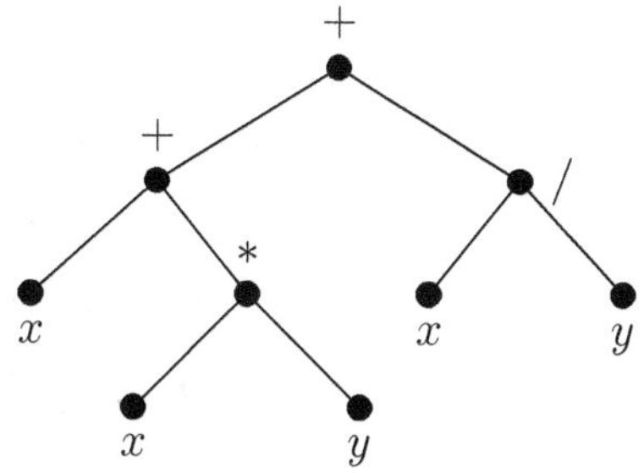

Figure 4.35

$$\text{Prefix notation}: \ + \ +x \ * \ x \ y \ / \ x \ y$$

$$\text{Postfix notation}: \ x \ y \ * \ + \ x \ y \ / \ +$$

$$\text{Infix notation}: \ (x \ + (x*y)) + (x/y).$$

Example 4.31. *What is the value of the prefix expression $* \ / \ 9 \ 3 \ + \ * \ 2 \ 4 \ - \ 7 \ 6$.*

Solution:

$$\text{Given expression is } * \ \underline{/ \ 9 \ 3} \ + \ * \ 2 \ 4 \ - \ 7 \ 6$$

$$= * \ 3 \ + \ * \ 2 \ 4 \ \underline{- \ 7 \ 6}$$

$$= * \ 3 \ + \ \underline{* \ 2 \ 4} \ 1$$

$$= * \ 3 \ \underline{+ \ 8 \ 1}$$

$$= * \ 3 \ 9$$

$$= 27.$$

Example 4.32. *Write the following expression in polish notation.*

$$(5x + 7y)^2.(x - 4)^3$$

Solution:

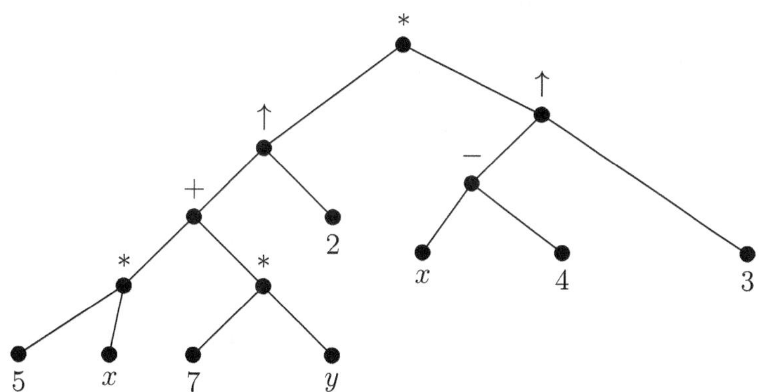

Example 4.33. *What is the value of the prefix expression* $+ - \uparrow 3\ 2 \uparrow 2\ 3\ /\ 6\ -\ 4\ 2.$

Solution:

$$\text{Given expression is } + - \uparrow 3\ 2 \uparrow 2\ 3\ /\ 6\ \underline{-\ 4\ 2}$$
$$= + - \uparrow 3\ 2 \uparrow 2\ 3\ \underline{/\ 6\ 2}$$
$$= + - \uparrow 3\ 2\underline{\uparrow\ 2\ 3}\ 3$$
$$= + - \underline{\uparrow\ 3\ 2}\ 8\ 3$$
$$= + - 9\ 8\ 3$$
$$= + 1\ 3$$
$$= 4.$$

Example 4.34. *What is the value of the postfix expression* $5\ 2\ 1\ -\ -\ 3\ 1\ 4\ +\ +\ *$

Solution:

$$\text{Given expression is }\quad 5\ \underline{2\ 1\ -}\ -\ 3\ 1\ 4\ +\ +\ *$$
$$= \underline{5\ 1\ -}\ 3\ 1\ 4\ +\ +\ *$$
$$= 4\ 3\ \underline{1\ 4\ +}\ +\ *$$
$$= 4\ \underline{3\ 5\ +}\ *$$
$$= 4\ 8\ *$$
$$= 32.$$

Exercise:4.2

1. Draw all non-isomorphic trees on 5 vertices.

2. If T is a tree on 10 vertices, then find total degree of T.

3. Define spanning tree. Draw all possible non isomorphic spanning trees of the following graph.

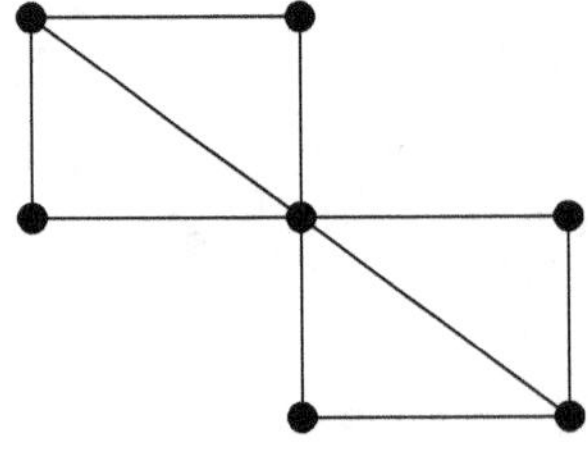

4. State true or false with justification: 'There exists a binary tree on 12 vertices'.

5. If T is a binary tree on 15 vertices then find the number of pendant and non pendant vertices in T.

6. Find center radius and diameter for the following trees.

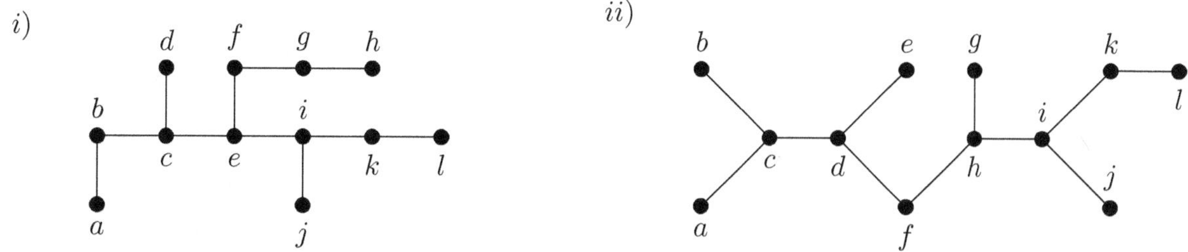

7. Using Kruskal's algorithm, find the shortest spanning tree of the following graph. Also find the weight of the shortest spanning tree.

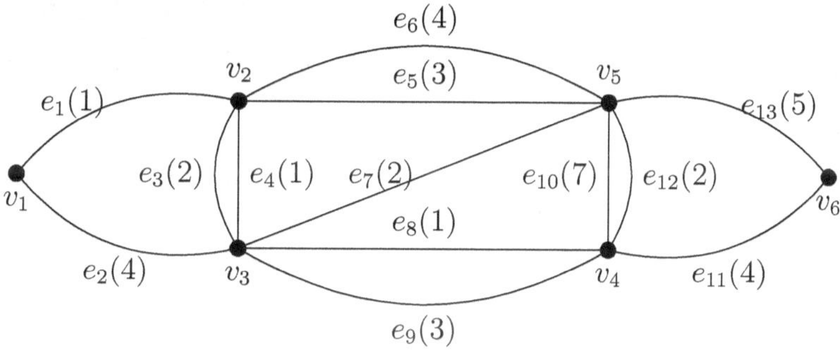

8. Use Kruskal's algorithm to find a minimal spanning tree in the following weighted graph.

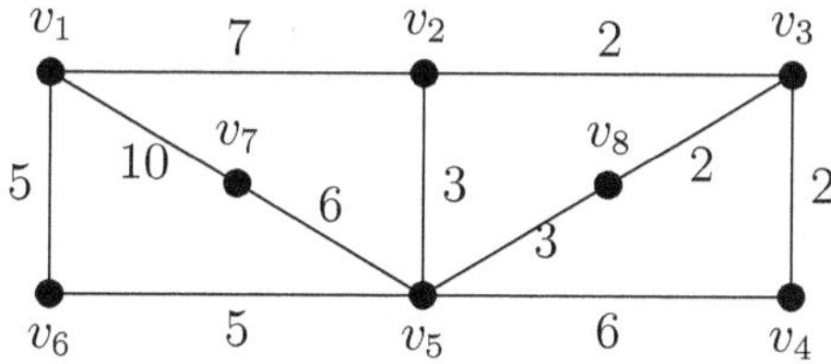

9. Find a shortest spanning tree using Prim's algorithm.

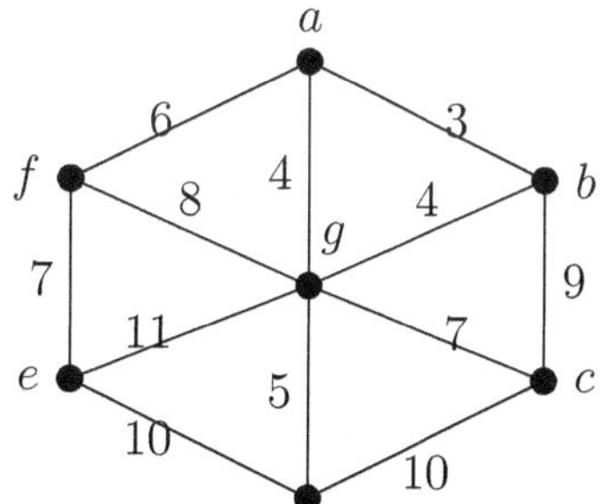

10. Draw a full 3-ary tree of height 2.

11. Consider the following tree.

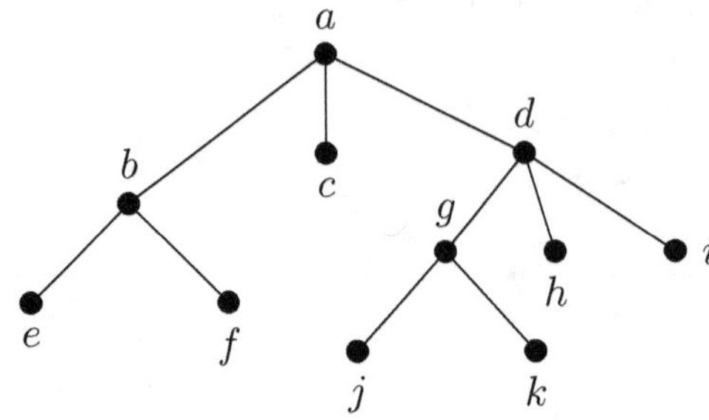

Figure 4.36

   i) Write the names of internal vertices.   ii) Write the names of children of d.

   iii) Write the names of parent of b.   iv) Draw subtree of vertex d.

12. A full 3-ary tree as 4 internal vertices then find total number of vertices in a tree. Also find number of leaves.

13. In which order does a preorder traversal visit the vertices in the following ordered rooted tree?

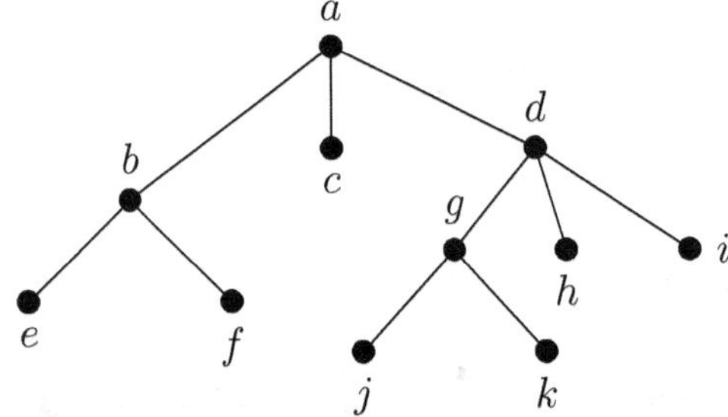

Figure 4.37

14. In which order does an inorder traversal visit the vertices in the following ordered rooted tree?

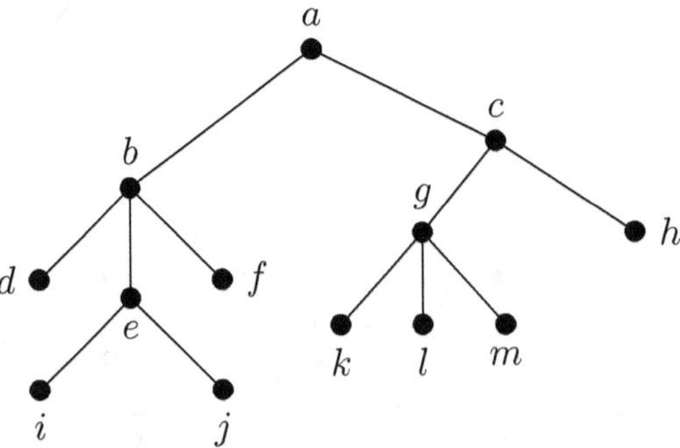

Figure 4.38

15. In which order does a post order traversal visit the vertices in the following ordered rooted tree?

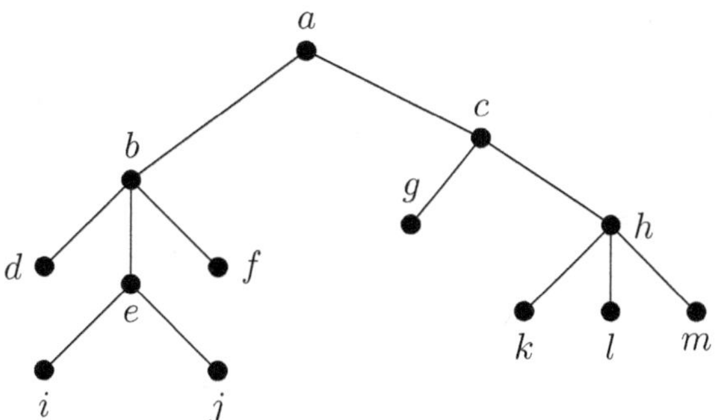

Figure 4.39

16. Draw the ordered rooted tree that represents the expression $(2a - b)^3(x + 4y)$. Also write the expression in polish notation (Prefix form)

17. Write the following expression in prefix as well as postfix notation: $\dfrac{(x - y)^2}{(3x + 5)^4}$

18. What is the value of the following prefix expressions?

    i $+ - * 2\,3\,5 / \uparrow 2\,3\,4$

    ii $* - * 2\,5\,3 / 4\,2$

    iii $\uparrow - * 3\,3 * 4\,2\,5$

19. What is the value of the following postfix expressions?

    i $7\,2\,3 * - 4 \uparrow 9\,3 / +$

    ii $2\,5 * 3 - 4\,2 / *$

    iii $3\,2 * 2 \uparrow 5\,3 - 8\,4 / * -$

Solutions

1.

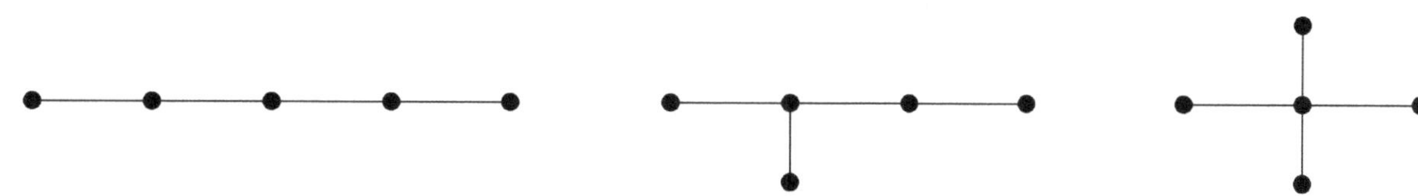

Figure 4.40

2. 18

3.

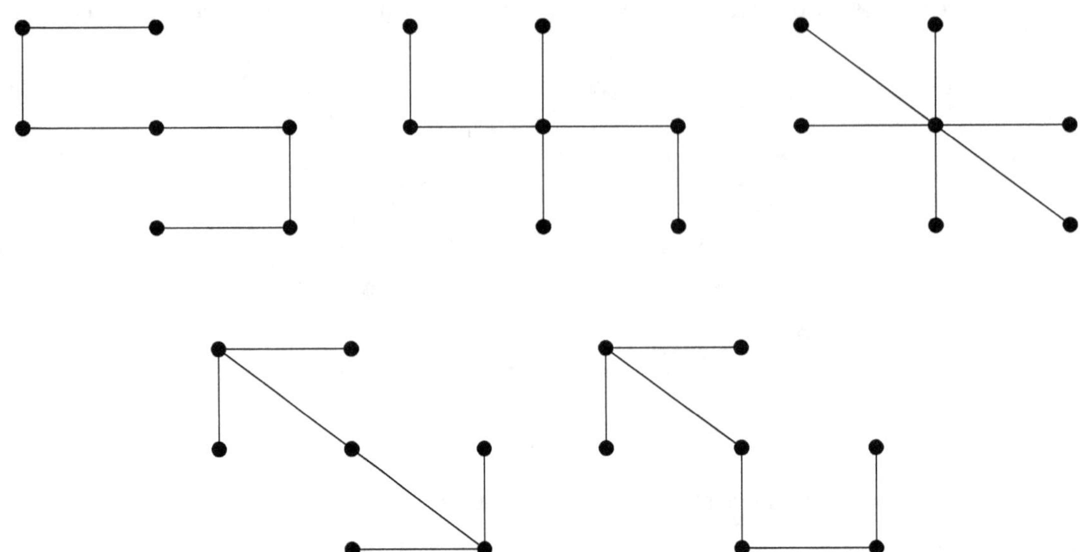

Figure 4.41

4. False (12 is an even number)

5. Pendant vertices $= \dfrac{n+1}{2} = 8$, Nonpendant vertices $= \dfrac{n-1}{2} = 7$.

6. i) center: e, radius=3, diameter=6.   ii) center : f and h, radius=4, diameter =7.

7. Shortest spanning tree has weight 9.

8. Weight of minimal spanning tree is 25.

9. Weight of shortest spanning tree is 32.

10.

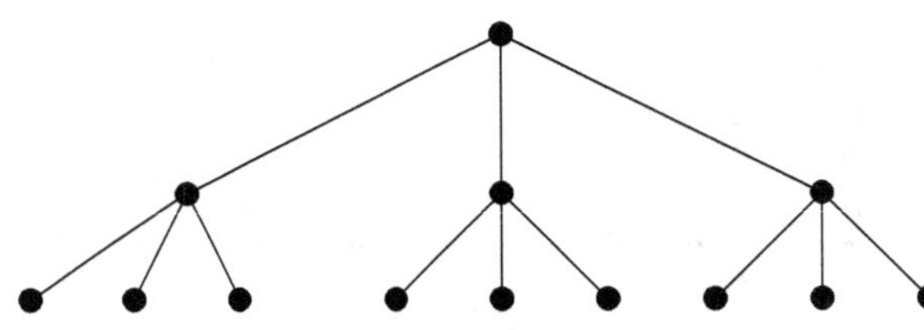

Figure 4.42

11.   i Internal vertices: a, b, d, j   ii.   Children of d: g, h, i

      iii Parent of b: a   iv.   Subtree at d.

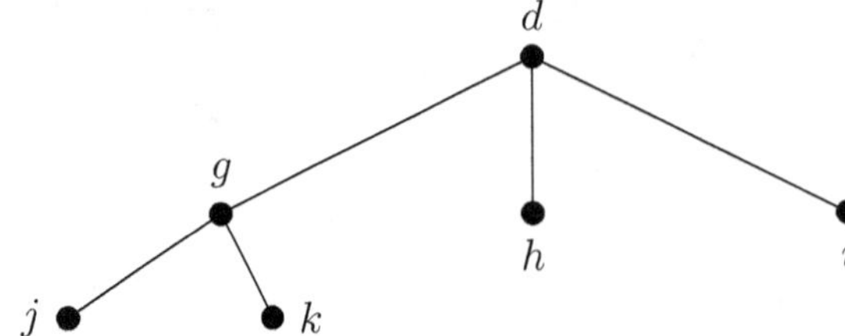

12. $n = m.i + 1 = 13, \quad l = \dfrac{(m-1)n+1}{m} = 9$

13. a b e f c d g j k h i

14. d b i e j f a k g l m c h

15. d i j e f b g k l m h c a

16.

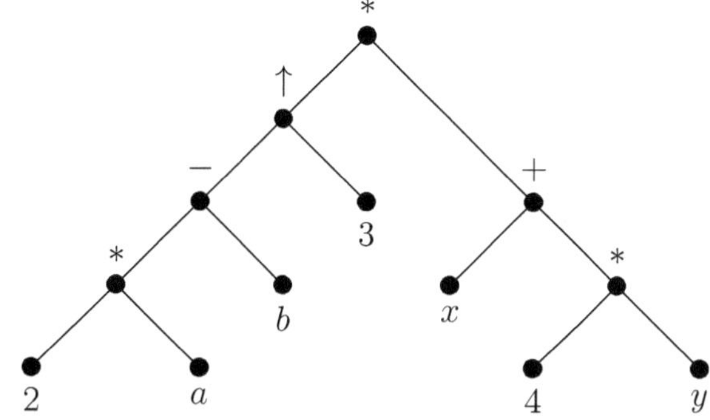

Polish notation $*\ \uparrow\ -\ *\ 2\ a\ b\ 3\ +\ x\ *\ 4\ y$

17.

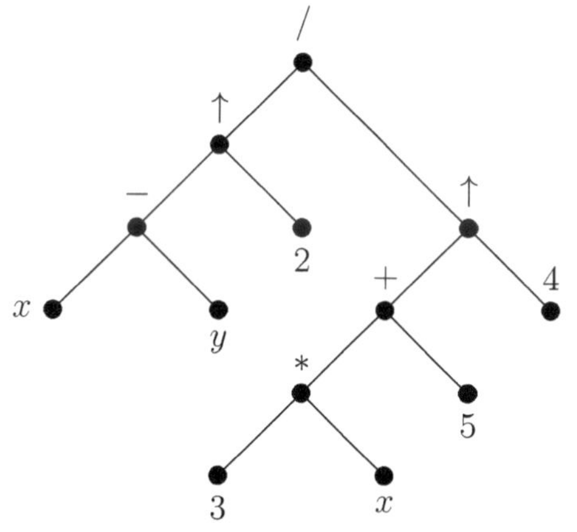

Prefix form : $/\ \uparrow\ -\ x\ y\ 2\ \uparrow\ +\ *\ 3\ x\ 5\ 4$

Postfix form : $x\ y\ -\ 2\ \uparrow\ 3\ x\ *\ 5\ +\ 4\ \uparrow\ /$

18. i) 3   ii) 14   iii) 1.

19. i) 4   ii) 14   iii) 32.